Energy-Saving Projects for the Home

Created and
designed by the
editorial staff of
ORTHO Books

Written by
Bill Henkin

Art Direction
and Design by
John Williams
Barbara Ziller

Drawings by
Ron Hildebrand

Final illustrations by
Ellen Blonder
Sally Selmeier
Shandis McKray

Photography by
Fred Lyon
Josephine Coatsworth
Mike Landis

Ortho Books

Publisher
Robert L. Iacopi

Editorial Director
Min S. Yee

Managing Editor
Anne Coolman

Horticultural Editor
Michael D. Smith

Production Editor
Barbara J. Ferguson

Editorial Assistant
Maureen V. Meehan

Administrative Assistant
Judith Pillon

Copyediting by
Editcetera
Berkeley, CA

Typography by
Terry Robinson & Co.
San Francisco, CA

Color Separations by
Colorscan
Palo Alto, CA

Address all inquiries to:
Ortho Books
Chevron Chemical Company
Consumer Products Division
575 Market Street
San Francisco, CA 94105

Printed in August, 1980

1 2 3 4 5 6 7 8 9 10

ISBN 0-917102-86-X

Library of Congress Catalog Card Number 80-66348

Acknowledgements

Special thanks to:

Pacific Gas & Electric Company
San Francisco, CA
David M. Williams,
Conservation Analyst
Richard A. Deus
William McKee
Billye A. Austin
Stephen A. Adams
Noelle M. Buer
Izetta Feeny

Joanne Sutro
Energy Support Team
San Francisco, CA

Editorial Consultants:

Jim Augustyn
Berkeley Solar Group
Berkeley, CA

Energy Resources Center
U.S. Department of Energy
San Francisco, CA

Molly Fleischner
Unicorn Realty
San Francisco, CA

William H. Grist
Atlanta Gas & Light
Atlanta, GA

Neal Jurgenson
Shrader Wood Stoves
San Rafael, CA

Wes Meyers
Barling-Meyers, Ltd.
Half Moon Bay, CA

Jim Phillips
Sears Service Center
San Francisco, CA

Arthur Rosenfeld
Lawrence Berkeley
Laboratory
Berkeley, CA

John A. Zizolfo
Consolidated Edison Co.
New York, NY

Photographic Consultants:

Automatic Solar Pool Covers
Concord, CA

David Baker
Sol-Ark
Berkeley, CA

Dan Beswick

Donald G. Boos
Landscape Architect
Palo Alto, CA

Interactive Resources, Inc.
Solar Energy Consultants
Point Richmond, CA
Tom Butt
Tom Campiglia
Burr Nash

Jacobson & Silversteen
Architects
Berkeley, CA

Rollamatic Roofs, Inc.
San Francisco, CA

Claude Stoeller
Stoeller Partners
Berkeley, CA

Sun Light & Power Co.
Point Richmond, CA
Gary Gerber
Ed Nold
Steve Spicer

Photography by:

(Names of photographers are followed by the page numbers on which their work appears)

Josephine Coatsworth: 10, 66, 67, 74, 75, 78, 79, 82, 83, 89, 91, 102

Mike Landis: 6, 8, 9, 14, 15, 16, 103

Fred Lyon: 4, 12, 20, 50, 68, 94

Front Cover:

After taking the proper steps to plug the energy leaks in your house, particularly insulating your attic, you might want to look into solar energy systems or wood-burning stoves.

Front cover photographs by: Fred Lyon
Back cover photograph by: Tom Tracy

Energy-Saving Projects for the Home

Whether we drill for oil, dig for coal, harvest the forests, split the atom, or harness the sun, there is no question that energy is going to cost us more in the near future than it used to. The watt you save may be your own dollar.

Saving Energy = Saving Money

Once upon a time, and not so very long ago, it seemed as if the fuels we used for energy would always be abundant and inexpensive. The United States had a wealth of its own energy resources and for many years was the world's foremost producer of oil and natural gas. Not until the middle of the 1950s did the lower prices of fossil fuels from other countries make importation an economic necessity.

In those early years we did what humans have always done: We used what the earth provided, first to fulfill our needs and then to fulfill our desires. As the leading industrial nation in the world we enjoyed a high standard of living, which contributed to, but was not the sole cause of, the fact that our energy consumption doubled in the 20 years following World War II; and that rate of growth continued to increase. To support our energy consumption we imported more foreign oil. With about 6 percent of the world's population, the United States consumed almost a third of the world's energy. We used some of that energy to produce about a quarter of the world's industrial and consumer goods, and about half of the world's food exports.

The Arab oil embargo raised the price of crude oil imported into this country from $3.41 per barrel in 1973 to $11.11 per barrel in 1974. That precipitous price rise was reflected most obviously at the gas pumps and in home heating fuel costs, and we began to speak about an "energy crisis." As fuel prices increased even more over the following years we were forced for the first time in modern history to examine our energy consumption and production patterns. We found that as a nation and as individuals we had become heavily dependent on, even addicted to, cheap fossil fuels.

Now, in the 1980s, oil fills more than half our nation's total energy requirements. Since we import more than half the oil we use—at an annual cost of some $90 billion—we find our country's economy dangerously vulnerable to interruptions in its oil supply, and equally vulnerable to foreign oil price hikes.

Meanwhile, despite frequent and substantial increases in the costs of our gas and oil, we continue to burn up these nonrenewable fuels at a prodigious rate. *Non renewable* means simply what it says: What we use is gone forever.

Reliable experts assure us that we are a long way from exhausting the world's reserves of any of the fossil fuels. But we have come close to exhausting the gas and oil supplies that are easy, and therefore inexpensive, to reach. Most new gas and oil explorations are taking place, not in the open fields of Texas and Oklahoma, but in the frigid waters of the North Sea, or four miles beneath the surface of Louisiana. If these explorations are successful, decades worth of **oil and gas** may become available to us.

Considerable amounts of **coal**—both soft bituminous and hard anthracite—remain unmined in many parts of the world. According to some experts, the United States alone possesses enough coal to supply our current home energy needs for several centuries, although we must find a way to keep the air clean if we are to use it.

Nuclear fission is an established source of energy throughout the world, and generates about 10 percent of the electrical energy used in the United States. It is to be hoped that the controversies that have surrounded its further development will be satisfactorily resolved. **Nuclear fusion** remains a potential energy source only, since it has not yet been clearly demonstrated to be a safe, controlled, and practical resource.

Geothermal energy, using the natural heat from inside the earth, is practical only where the underground magma nears the planet's surface. For this and other reasons geothermal will provide energy in the future, mostly in the form of electrical generation.

Electricity, which must be produced by converting other sources of energy, is generated by coal, oil, gas, nuclear power, and water power. **Hydroelectric** plants will contribute to that generation, as they have for years; and the wind is being explored as another clean, renewable, and easily converted energy source.

Laboratory chemists are developing synthetic oil and gas from coal, shale rock and tar sands. These **synthetic fuels** are already in production in several American plants. But they are still somewhat experimental, and even if the attendant problems are solved in the very near future, we are at least several years away from having enough volume of such fuels to make a major contribution to our energy needs.

Solar energy, the focus of intense development in the past decade, has demonstrated some useful capabilities, particularly in its simple, passive forms.

Every one of these sources of energy is likely to be applied in the near future, in the United States and elsewhere in the world. We have only recently agreed that there is an energy problem; and we are still seeking the solutions that will carry us through the long term.

Increased exploration and production costs result, of course, in higher fuel costs to the nation and to the individual. While we are not in fact on the verge of running out of energy sources, the abrupt difficulty and expense of securing what we used to get so easily and cheaply suggest that we had best take some action to ease the strain on ourselves for the next few decades.

Obviously, most of us are not about to grab our drilling rigs and go off in search of new oil wells. But surprisingly, each of us can uncover some very substantial fuel reserves right in our own homes. We can start to ease the energy burden on our nation's economy, and simultaneously ease the strain on our own wallets and checkbooks.

That is what this book is all about: reducing your energy expenditures in ways your can see and measure. In a very real sense, saving energy has come to equal saving money. As individuals, we do not have to wait for national policy before we take action on our own behalf. We can make the difference in our own energy bills. And when enough of us do so we will make a collective difference in the self-sufficiency of our country.

Energy-Saving Projects for the Home is made up of six chapters. Rather than exploring geothermal energy, or nuclear plant development, these pages are designed to be of practical use to you. The book is intended to present you with options, so that you will know what you can do to reduce your energy consumption and therefore your reliance on our increasingly expensive supplies of oil and gas.

It's so easy to overlook small cracks in your house—some are almost impossible to see without really looking for them. But these cracks are the "holes" through which your heat—and your energy dollars—slip out. See pages 28 to 31 for details on caulking all around your house.

You will find that your options range from extremely simple steps that cost you nothing yet bring you measurable savings within 30 days, to complex installations that may not pay back for years, but that will immediately add to your comfort and energy-independence.

The chapters that follow are intended to be guidelines, and not cookbook solutions to energy questions. Because every situation is different, any given form of energy efficiency, from a simple conservation measure to an extensive solar installation, may or may not be worthwhile in your particular case. This book should help you come to your own decisions.

Conservation

Of all the options open to you, conserving what you already use—gas or oil or electricity—is by far the easiest, fastest, cheapest, and most direct and cost-effective way to save energy and energy dollars. There is no debate on this issue. But before you envision yourself and your family shivering and huddled around a fire, let us reassure you that conservation does not mean going without comfort. Conservation is about the *quality* of energy use, not the quantity. It only appears to be about quantity because when energy is used wisely, less of it is used.

Residential buildings use about 20 percent of all the energy consumed in the United States, and of that about 70 percent is used for space conditioning: the heating and cooling of living areas. Yet only about 2 percent of American homes are insulated well enough to retain their heat in cold weather, or their coolness in hot. You would be horrified if it were suggested that you leave your front door open all winter and let your heating dollars fly away. But according to the U.S. Department of Energy, a 1/4 inch thick crack under your front door will waste about as much of your heating bill as if you had a 2- by 2-inch hole in your wall. You may also be losing heat through the cracks above and below your exterior doors; the tiny crack around your mail slot; the little chink out around the exhaust duct of your clothes dryer; a few loose windows . . . it all adds up. You can see that it's worth your while to slow the rate at which your house loses the air you have paid to heat or cool to your liking.

For maximum long-term home energy conservation gains, a general program that covers all your energy drains will probably serve you best. Only you can design that program, because only you can find out exactly where and how much of your energy is being wasted.

Perhaps you don't want to invest a great deal of time and effort in a conservation program just yet. You might prefer to get started in a small way, to see what conservation feels like in your life. In the pages that follow, you will find some no-cost common sense ways to reduce your energy consumption immediately. By paying attention to your daily energy-consuming patterns, you can make small changes that won't cost you a penny but could easily save you 10 to 20 percent of your total energy bill.

If you choose to make larger changes around the house, you may find financial assistance just a phone call away. Before you make any investment in an energy-saving measure, you should check out the various sources for such financial help. The last two pages of this chapter outline some of the options open to you, including low-interest loans and tax incentives.

Whether you choose to make big changes or small ones, you can't really know what you are wasting until you know what you are spending, and before you make *any* changes, you should know where *your* house is leaking the most energy. The next chapter, "The Energy Audit," presents a home audit that you can do yourself. This audit will give you a good idea of where you can take steps to save energy and energy dollars that will be the most cost-effective for your particular situation.

The following chapter, "Plugging the Energy Leaks: The House," takes a close look at the house, describing in detail some of the alterations you might have to make in its structure that will save energy. We also examine the costs of these alterations, so that you can begin to judge whether they are worthwhile *for you.*

In the third chapter, "Plugging the Energy Leaks: The Systems," we look into the systems that operate in your house to keep it warm or cool, examining the changes you can make to save money and energy. And again we offer some guides that will enable you to determine whether these are changes you care to make.

After examining your house as it stands now, the next chapter, "Solar Energy," introduces the theory and practice of solar energy applications. Can the sun solve *your* energy problems? Only you and the shadows can know for sure. In the final chapter, "Wood and the Other Alternatives," instead of looking resolutely forward, we look back at an energy source that supplied virtually 100 percent of the fuel energy used by our forefathers when they came to this land: wood. But something has changed about the use of fuel wood in the past couple hundred years, and that is the firebox it's burned in. Therefore we will look at some of the different kinds of stoves and fireplace amendments you can buy that will make burning wood an efficient and heart-warming experience.

And finally, we will take a quick look at wind and water as sources of home energy production. While these are still a long way from being in common use by the average homeowner, more and more people are investigating them as alternatives.

No matter what you do, the first step is to increase your awareness of how much money you are throwing away by continuing your old, familiar energy-consuming habits. We all have them, but changing these habits doesn't have to be painful. It can be a pioneering adventure into the energy-saving patterns of the future.

No-cost, No-work Energy-Saving Ideas

Conserving energy is a little bit like going swimming. People eager to practice their kicks may leap right into the deepest waters, while others, less certain of the process or the results they may expect, dip a tentative toe in the shallows and wait a moment to see what happens before committing themselves further. But what all swimmers have in common is the need to get wet before taking the first real stroke.

In the next couple of pages we want to remind you that getting wet can be simple and painless. Whether you are already serious about saving energy, or just beginning to think it might be a good idea, the advice that follows should be the basis of your must-do list. These are some of the easiest, most common-sense *free* steps you can take to reduce your home energy consumption and save your first home energy dollars. All it takes is a small change of habits.

It has been estimated by the Department of Energy and other expert watchdogs of our nation's resources that a household may save as much as 10 to 20 percent on home energy bills by following just such steps as these. Of course, we cannot say exactly how much taking these measures will save *you,* since the results you achieve will depend on such variables as where you live, how energy-conscious you are already, and the particular necessities of your life-style. If, for example, you are unwilling to give up the luxury of standing in the hot shower for ten minutes every morning, the bill for fuel to heat your water will not drop as precipitously as it might if you learned to get in, get wet, turn off the water, soap up, turn on the water, rinse off, and get out. But there are measurable savings in every item on this list—and you will undoubtedly be able to add some suggestions of your own.

■ **Refrigerator:** Keep condenser coils clean for greater efficiency.

Turn thermostat setting up from 37°F to 40°F.

Turn off ice maker when not in use.

Keep refrigerator horizontally level for efficient operation (a glass of water on a refrigerator shelf will show whether it's level).

Defrost when frost is 1/4-inch thick.

Keep doors tightly closed.

Open doors only when necessary.

Close doors as soon as you've put food in or taken it out.

Cool hot foods before refrigerating.

Organize foods for ease in locating to minimize time with door open.

Cover all liquids—high humidity requires more energy.

■ **Stove:** If you buy a gas range, select one with a pilotless ignition.

Check gas stove flame color: yellow traces mean burners may be clogged and require cleaning.

Self-cleaning ovens are huge energy consumers; clean them by hand.

Defrost foods before cooking.

Preheat your oven *only* long enough for it to reach the desired temperature.

Match pots and pans to burner size.

Use flat-bottomed pans with tight-fitting lids.

Keep oven doors tightly closed.

Use surface units when possible; ovens and broilers require more energy.

Make one-pot meals such as stews and casseroles.

Bake several things at once instead of a single item—the oven is going to heat up anyway.

Do not peek into the oven while it's on—each peek drops the temperature by 25°F to 50°F or even more.

Use toaster ovens or portable appliances for small quantities of food.

Use microwave ovens to cook small quantities of food. (Microwave ovens can be energy wasters if used to cook for five or more people.)

Heat a ceramic tile while baking. Use the tile rather than the oven to keep rolls hot.

Warm dishes or breads in retained heat after oven is turned off.

■ **Dishwasher:** Rinse dishes in cold water before placing them in the dishwasher.

Keep dishwasher drains and filters clear of debris.

Run dishwasher for full loads only.

■ **Laundry:** Rinse clothes in cold or warm water, not hot.

Wash full loads only, but remember that overloading wastes energy, too.

Special features on the washer, such as soak cycles and suds savers, conserve water and cut down on operating time of the machine—use whenever possible.

Use a clothesline instead of dryer whenever possible.

Clean lint filter on dryer after every use.

Overdrying articles makes clothes harsh-feeling and wastes energy.

Iron as little as possible—ironing uses as much energy as ten 100-watt bulbs. Instead, remove clothes from dryer immediately so that wrinkles do not set in.

■ **Bathroom:** Turn temperature of water heater down to 120°F.

Fix leaking faucets and worn washers immediately.

Use flow restrictors in sink and shower.

Do not run hot water continuously while washing hands or shaving.

Take showers—they generally use less hot water than baths.

Take shorter and cooler showers.

■ **Lights:** Turn off lamps when not in use.

Use the lowest setting on 3-way bulbs whenever possible, particularly when watching TV.

Use lampshades with a white liner for better reflection.

Place work tables and desks near windows to take advantage of daylight.

Use fluorescent fixtures whenever possible.

Arrange lamps to reflect off two walls—more light is reflected than if the lamp is against one wall.

■ **Heating:** Winter thermostat settings: daytime, 65°F to 68°F; nighttime, 60°F; unused areas, 50°F.

Keep windows clean to let in more sun.

Always keep heating sources clean. Dirty filters, grills, and coils mean wasted energy.

Keep chimney clean.

Completely seal off any unused doors to the outside for the winter (submitted by C. L. Miller).*

Keep all exterior doors closed tightly.

Keep mail chutes and milk chutes closed.

Close damper when the fireplace is not in use.

Use shades, drapes, shutters or curtains on all windows—they slow heat loss through glass.

Keep drapes open on sunny days and closed on cloudy days and at night.

Do not heat unused rooms, and keep doors to these rooms closed.

Keep furnace cold-air register unobstructed.

Close windows near thermostat.

Keep indoor/outdoor traffic to a minimum.

Move furniture that blocks hot- or cold-air registers.

Lower the thermostat if a large group of people will be in the room—their body heat will warm the room.

Wear extra layers of clothing for warmth.

Do baking in afternoon and keep small children in the kitchen where it's warm (submitted by Bonnie Allen).

Never use the oven just to heat a room, but do take advantage of the heat when you must use the oven.

After baths are finished, allow the steam to heat the bathroom and upstairs rooms. Drain water when cold (submitted by Bonnie Allen).

Close bathroom door at night before turning down thermostat, keeping heat in bathroom (submitted by Mrs. Leo Nothman).

If you sleep with windows open, which wastes energy, at least shut the bedroom registers and close the door.

If the outdoor temperature is 30°F lower than indoor temperature, a fire in the fireplace can draw warmth from the house and result in a net heat loss! Reduce thermostat setting and close all doors to the room with the fire.

Burn only dry wood in the fireplace.

* At our invitation, energy-saving ideas were submitted by Ortho Books readers. Winners will receive $25, and their names appear immediately after their ideas.

Windows are essentially large holes in the walls of your house, and not very well-covered ones, at that. This automated shutter can let the sun in or keep it out, and it can be closed at night to insulate against nighttime heat loss. Various window treatments are covered on pages 32 to 37.

■ **Cooling:** Keep drapes closed on sunny days and open on cloudy days and at night during the summer months.

Do not run a humidifier when the air conditioning is operating.

Protect cooling unit from the sun and from plants that restrict air flow around it.

Close heating vents when using room air conditioner so that the vents do not fill with cold air.

■ **Miscellaneous:** Unplug "instant on" feature of the TV—it is a constant drain of electricity.

Turn off appliances when they are not in use—television when no one's watching, radio when no one's listening.

If you have a waterbed, keep it covered to retain heat.

Put new seedlings on top of refrigerator or hot water heater for bottom heat (submitted by Bob Rutemoeller).

Eliminate outdoor decorative lighting.

■ **Vacation:** Be sure all faucets are turned off and do not leak.

On central furnaces with a disconnect switch, flip the switch to OFF.

If power supply to any electrical equipment is a simple plug, pull the plug.

Turn water heater gas control valve to "pilot."

Use up perishables and then set refrigerator to warmest setting (not defrost).

Turn off swimming pool heater and pilot and reduce time clock so filter operates less.

If you'll be gone for longer than a month: Turn off air conditioner. Turn off heat and pilot light. Empty refrigerator, turn off, and leave door ajar.

State Tax Incentives

Before you begin any conservation project, you should explore every possible source of financial assistance. Federal and state governments offer tax incentives that may cut your costs significantly. Some utility companies offer loans at low or no interest for energy conservation improvements, so you should be sure to contact them.

There are two distinct categories for financial assistance: weatherization, such as weatherstripping and insulation; and renewable energy resource systems, including solar, wind and geothermal energy.

In at least one state, Alabama, there is also a tax deduction if the primary source of heating in a residence is converted to a wood stove. While most states do not presently have tax incentives for wood stove heating in the

In many ways it seems as if a house is not a home without a fireplace. And yet most fireplaces are huge energy wasters. If you don't want to give up using your fireplace, there are some things you can do to increase its efficiency. Our last chapter, starting on page 95 will give you some ideas.

home, it's worthwhile to keep current with your state's programs for this area of energy savings. If you've been confused by conflicting reports about the kinds of savings, credits, deductions, and other incentives available for your home energy conservation projects, you are not alone. Sometimes even the people making the reports don't have the most recent data. The entire field of energy conservation is changing and growing so rapidly that Monday's brilliant new idea is often Tuesday's ancient relic. It's almost impossible to keep up with all the new laws, products, and data that could make a difference to you.

As of 1980 almost every state in the union offers *some* kind of tax incentive or loan program for the purchase, installation, and/or use of energy-saving measures. Those states that do not presently provide such assistance hope to develop programs shortly.

Because every state's program is different from every other one, and because many of these programs are being changed even as you read this page, the following information is designed to help you find the information you need.

Here are some definitions of common terms that will help you understand what is being offered to you.

■ **Credits:** A tax credit is the amount of money the government allows you to subtract from your owed income tax because of money you invested in energy conservation for your home. If the credit you have earned exceeds the taxes you owe, that does not mean you can expect a refund. However, the excess can be carried over and applied to the following year's taxes. This is not the case with most other tax credit situations.

■ **Deductions:** Deductions are also allowed for money you have invested in energy conservation, but these are subtracted from your total income even before you start to figure your taxes. They actually lower your real, taxable income.

■ **Loans:** In many states, utility companies and related agencies offer conservation loans to their customers at very low interest rates, and sometimes with no interest at all. So if you are considering an investment in an energy-saving project, be sure to call your utility company to find out what financial assistance programs they may have.

■ **Loans to Veterans:** Some states will increase their usual loan limit to veterans who use approved energy-conserving methods in their homes.

■ **Property Tax Exemptions:** In many states, if you increase the value of your house by adding energy-conserving devices to it, your property tax cannot be increased on that basis. For example, in the state of Washington, until the year 1986, solar power systems and solar water and space heating systems will not increase the assessed and taxable values for property tax.

■ **Sales Tax Exemptions:** In some states the sale of energy-conserving equipment cannot be taxed.

■ **State Tax Credit:** In some states, the amount of credit you can claim for energy conservation measures is reduced by the amount of such credit you claim from the federal government.

To be eligible for state tax credit, systems are expected to last for three years.

The list on page 11 gives addresses and phone numbers of the main energy department office in every state. As with everything else in the energy field, it is difficult to stay current because things are moving so fast. By the time you write or phone your state's office, it may have moved. At the very least however, these numbers will lead you to the numbers of the agencies that *can* provide you with the information you need. These agencies are the best places to start; if they can't help you, they will be able to tell you who to call next.

Federal Tax Incentives

Federal weatherization credits cover insulation, weatherstripping, caulking, storm windows, set-back thermostats, pilotless ignition added to a furnace, and a damper added to a flue. The credit is 15 percent of the cost for the first $2000 spent. The maximum credit is $300.

Until January 1, 1986, you may subtract from your federal income tax 40 percent of the cost of any eligible *active* solar system or wind or geothermal installation purchased on or after January 1, 1980. The credit applies only to the first $10,000 you spend, for a maximum incentive of $4000.

Systems purchased before 1980 are subject to the old tax credits: 30 percent of the first $2000 spent and 20 percent of the next $8000, with a maximum of $2200 credit available.

The systems covered in these tax incentives are *active* systems and must be reasonably expected to last for five years after their installation. Items such as windows, extra-thick walls, skylights, greenhouses, and roof overhangs are not included as equipment because they serve other structural functions in the home and are considered *passive* forms of energy conservation.

Like state tax credits, federal tax credits will not lead to refunds; but any credit left over after all taxes have been deducted may be carried over onto the following year's tax form.

Figuring Your Finances

Solar collectors are becoming a more familiar sight. Solar energy systems are really quite simple in principle: they collect, store, and distribute the heat from the sun's energy in a controlled fashion. See pages 69 to 93 for descriptions and illustrations of these systems.

The following example is a step-by-step approach to figuring the cost of adding attic insulation, weatherstripping and caulking, and installing a solar hot water system in a four-bedroom home. This fictitious home is 2,400 square feet and arbitrarily located in Sacramento, California.

1. After deciding on the energy conservation methods listed above, contact your local utility company. For instance, in northern California the Pacific Gas and Electric Company provides a free home energy audit for its customers. They will show you where your home's energy leaks are and advise you on any aspects of your proposed project. They will also provide a list of local contractors to do the work for you.

2. List the cost of each project. The costs for all these projects can vary greatly, depending on how energy efficient you want your home to be. For example:

Solar water heater, contractor installed and certified	$2500
Attic insulation, contractor installed	$ 550
Latex acrylic caulking, used around entire home (about 20 tubes)	$ 45
Weatherstripping, foam-filled gasket stripping around all windows and doors	$ 135
Total	$3230

3. Check all possible sources for loans. In California, Pacific Gas & Electric provides a loan of up to $500 for attic insulation. The loan can be made for six months to five years at 8 percent interest, paid monthly beginning on receipt of the first bill. (A more extensive loan service is being approved right now.) So, in our example, the project will cost $2730 out-of-pocket.

4. Check out federal and state tax incentives. In this example, the credits allowed on federal and state taxes are:

■ **Federal:** For the insulation, weatherstripping, and caulking a taxpayer is allowed 15 percent credit for the first $2000 spent. The maximum credit is $300 (page 9). In our example, insulation, weatherstripping and caulking come to a total of $730. Therefore, the federal weatherization credit allowance is $109.50. Federal tax credit for a solar water heater is 40 percent of the first $10,000 spent, with a maximum of $4,000 credit. So, in this situation, the credit would be 40 percent of $2500, or $1000. That means that the total federal credits would be:

Weatherization credit	$ 109.50
Solar credit	$1000.00
Total federal credit	$1109.50

■ **State:** California offers no credits for weatherization, but a passive *or* active certified solar system is allowed 55 percent credit, with a maximum credit of $3000. For the solar water heater, 55 percent of $2500 equals $1375.

Since the California state tax credit is reduced by the *allowable* federal credit, whether or not the federal credit is taken, the state solar credit of $1375 would be reduced by the federal solar credit ($1000) for a total allowable state credit of $375. Therefore, the combined federal and state credits would be:

Federal Credits	
Weatherization	$ 109.50
Solar	$1000.00
State Credits	
Solar	$ 375.00
Total Credits	$1484.50

5. Figure out your actual costs and initial cash outlays. Of the total costs for all improvements in our example, with or without a loan, $1484.50 can be credited against your federal and state taxes.

This means that your total cost for the purchase and installation of your improvements would be $3230 less $1484.50, or $1745.50. If you do take a loan to reduce your initial cash outlay, you would add to these costs the additional interest on the loan. But, as noted above, the initial cash outlay would be reduced from $3230 to $2730.

6. Determine your expected savings from these improvements. (See the chart on page 17, and, for a sample of solar water heater savings, see page 89.) The combination of these figures—the cost of purchase and installation, estimated fuel savings from the improvements, term and amount of your loan repayment—will give you a relatively accurate picture of your energy dollar outlays and savings. With these kinds of calculations, you can decide which of the improvements seem most cost-effective for you.

7. Examine your alternatives. In our example, you might decide to complete only the weatherization part of your program. In that case, your total estimated cost would be $730. Of that total, $109 could be credited against your federal taxes, bringing your actual cost down to $621. And, with a $500 loan from your utility company, your initial cash outlay would be only $230. Your loan repayment could be as low as $10 per month, and your savings could be as much as 25 to 30 percent of your monthly fuel bill.

This example can guide you through the same steps for your personal situation in your state. You will not necessarily follow it exactly—don't be limited by it, or expect that your state or utility company will have the same programs and incentives. But be sure to check out all the options of financial assistance that may be available to you.

State Energy Information

National Solar Heating and Cooling Information Center:
800-523-2929

IRS Information
800-772-2345

Alabama
Development Office
Office of Governor
State Office Building
Montgomery, AL 36104
(205) 832-6960

Alaska
Division of Energy and Power Development
Department of Commerce
338 Denali Street
Anchorage, AK 99501
(907) 276-0508

Arizona
Energy Programs Office
Office of Economic Planning and Development
1700 W. Washington
Phoenix, AZ 85007
(602) 255-3303

Arkansas
Arkansas Energy Conservation and Policy Office
960 Plaza West Building
Lee & McKinley Streets
Little Rock, AR 72205
(501) 371-1379

California
Energy Resources Conservation and Development Commission
Resources Agency
704 11th & L Building
Sacramento, CA 95814
(916) 920-6811

Colorado
Office of Energy Conservation
1600 Downing
Denver, CO 80218
(303) 839-2507

Connecticut
Office of Planning and Management
Energy Division
20 Grand Street
Hartford, CT 06115
(203) 566-2800

Delaware
Delaware Energy Office
P.O. Box 1401
56 The Green
Dover, DE 19901
(302) 736-5647

Florida
State Energy Office of Florida
301 Bryant Building
Tallahassee, FL 32301
(904) 488-6764

Georgia
Office of Energy Resources
270 Washington Street, S.W.
Atlanta, GA 30334
(404) 656-5176

Hawaii
Energy Management and Conservation Office
Department of Planning and Economic Development
1164 Bishop Street, Suite 1515
Honolulu, HI 96813
(808) 548-4090

Idaho
Office of Energy
State House
Boise, ID 83720
(208) 334-3800

Illinois
Institute for Environmental Quality
222 South College
Springfield, IL 62706
(217) 785-2800

Indiana
Indiana Energy Office
Department of Commerce, Seventh Floor
Consolidated Building
Indianapolis, IN 46204
(317) 232-8940

Iowa
Iowa Energy Policy Council
215 East 7th Street
Des Moines, IA 50309
(515) 281-6679

Kansas
Kansas Energy Office
503 Kansas Avenue
Topeka, KS 66603
(913) 296-2910

Kentucky
Kentucky Department of Energy
P.O. Box 11888
Iron Works Pike
Lexington, KY 40578
(606 252-5535)
1-800-432-9014

Louisiana
Office of Conservation
Department of Natural Resources
P.O. Box 44275
Baton Rouge, LA 70804
(504) 342-5540

Maine
Office of Energy Resources
55 Capitol Street
Augusta, ME 04330
(207) 289-3811

Maryland
Maryland Energy Policy Office
Room 1302, State Office Building
301 West Preston Street
Baltimore, MD 21201
(301) 383-6810

Massachusetts
State Energy Office
73 Tremont Street
Room 700
Boston, MA 02108
(617) 727-1250

Michigan
Michigan Energy Administration
Department of Commerce
P.O. Box 30004
Lansing, MI 48909
(517) 373-0480

Minnesota
Minnesota Energy Agency
American Center Building
150 E. Kellog Blvd.
St. Paul, MN 55101
(612) 296-6720

Mississippi
Mississippi Fuel and Energy Management Commission
Suite 228, Barefield Complex
455 North Lamar Street
Jackson, MS 39201
(601) 961-5099

Missouri
Missouri Energy Program
Department of Natural Resources
Box 176
Jefferson City, MO 65102
(314) 751-4000

Montana
Energy Division, Department of Natural Resources and Conservation
32 South Ewing
Helena, MT 59601
(406) 449-3780

Nebraska
Nebraska State Energy Office
State Office Building
P.O. Box 94841
Lincoln, NE 68509
(402) 471-2867

Nevada
Department of Energy
1050 East Williams Street
Carson City, NE 89701
(702) 885-4840

New Hampshire
Governor's Council on Energy
26 Pleasant Street
Concord, NH 03301
(603) 271-2711

New Jersey
New Jersey Department of Energy
101 Commerce Street
Newark, NJ 07102
(201) 648-3290

New Mexico
Energy and Minerals Department
P.O. Box 2770
Santa Fe, NM 87501
(505) 827-2471

New York
New York State Energy Office
Agency Building 2
Empire State Plaza
Albany, NY 12223
(513) 474-2121

North Carolina
Department of Commerce
Energy Division
215 East Lane Street
Raleigh, NC 27611
(919) 733-2230

North Dakota
North Dakota Office of Energy Management and Conservation
1533 North 12 Street
Bismarck, ND 58501
(701) 224-2250

Ohio
Ohio Department of Energy
30 E. Broad Street, 14th Floor
Columbus, OH 43215
(614) 466-8476

Oklahoma
Oklahoma Department of Energy
4400 N. Lincoln Blvd.
Oklahoma City, OK 73105
(405) 521-2995

Oregon
Department of Energy, Room 111
Labor and Industry Building
Salem, OR 97310
(503) 378-4128

Pennsylvania
Governor's Energy Council
1625 North Front Street
Harrisburg, PA 17102
(717) 783-8610

Rhode Island
Rhode Island Energy Office
80 Dean Street
Providence, RI 02903
(401) 277-3370, or
(401) 277-3773, Collect

South Carolina
Office of Energy Resources
S.C.N. Building
1122 Lady Street
Columbia, SC 29201
(803) 758-8110

South Dakota
Office of Energy Policy
Department of Executive Management
Foss Building
Pierre, SD 57501
(605) 773-3603

Tennessee
Tennessee Energy Authority
250 Capitol Hill Bldg.
Nashville, TN 37219
(615) 741-3023

Texas
Texas Railroad Commission
P.O. Box 12967, Capitol Station
Houston, TX 78711
(512) 475-0510

Utah
Energy Office
Department of Natural Resources
231 East 200 South
Salt Lake City, UT 84111
(801) 533-5424

Vermont
Vermont State Energy Office
State Office Building
Montpelier, VT 05602
(802) 828-2393

Virginia
Office of Emergency and Energy Services
310 Turner Road
Richmond, VA 23219
(804) 745-3245

Washington
State Energy Office
400 East Union
Olympia, WA 98504
(206) 753-4409

West Virginia
Fuel and Energy Division
West Virginia Office of Economic and Community Development
1262½ Greenbrier Street
Charleston, WV 25311
(304) 348-8860

Wisconsin
Office of State Planning and Energy
1 West Wilson, Room B 130
P.O. Box 511
Madison, WI 53701
(608) 266-8234

Wyoming
Energy Conservation Office
Capitol Hill Office Bldg.
320 West 25 Street
Cheyenne, WY 82002
(307) 777-7131

INCHES
ENGINEERS
LOCK JOINTS
MADE IN U.S.A.

Until you know how much energy you're using now, you can't know how much you can save. And until you know where you're spending it, you can't know where to save it. And the easiest way to find out won't cost you anything.

The Energy Audit

How to do a Home Energy Audit

Your house is about to be audited—not by the IRS, by *you*. Doing a walk-through energy inspection of your home will give you a good idea of the shape it's in, energy-wise. You'll be looking around, taking notes, checking off answers to questions, and even scribbling a few drawings, where appropriate. This audit will show you what your house needs, and where. Later, you can use the material in the rest of this book to sort out your priorities about energy-saving amendments, and then install (or have installed) the ones you choose.

To find out how well your home is being heated or cooled, get out a pen, paper, a clipboard, and this book. (Before you actually begin your walk, read through this section, including "A Simple Checklist" on pages 18-19.) You'll need to make a map of your house. You can use plain paper or graph paper. If you use the latter, let one small square equal one foot of your floorspace (two feet if your house is very large). Measure the length, width, and height of each room. Include every wall, door, window, and special feature of each area you'll be walking through—the outside; the general living space (living room, bedrooms, kitchen, bathroom, and so forth); the attic; and the basement.

This audit will take place outside, inside, and in-between. Carry your map with you, and make notes and brief sketches on it in the appropriate places.

Outside

■ **Cracks and Joints:** Walk completely around your house and look for cracks in the wall or joints. Pay special attention wherever two different kinds of building materials meet—where a concrete foundation joins wooden or aluminum siding; where a window or a door frame fits into a wall; where the roof joins the siding; and especially where the siding meets a chimney. If there are holes meant to accommodate wiring, pipes, ductwork, tubes, or hoses, check to make sure the holes aren't larger than they should be. These cracks should be filled with caulk to eliminate energy leaks through them. In some cases, they may have been caulked before. Check to see if the caulking needs to be replaced or if the seal is still intact. Complete instructions for caulking can be found on pages 28-31.

■ **Landscaping:** Surprisingly, what's outside your house affects your comfort inside it. Are you on a hill or slope, protected from wind or in its direct path? Do you have mature trees on your property? If so, how tall are they? Where are the trees located? Do they block northerly or southerly winds? How far away from the house are they? Do you have immature trees that will grow? To what height? Are the trees deciduous (lose their leaves in spring) or evergreen (keep their leaves year round)?

Is there shrubbery next to the house? Are there vines on or next to the house? Note all these details, sketching the greenery in on your map. Do you have places that are particularly battered by winds—doorways, or one side of the house? Do you have fences anywhere? How high are they and where are they located? What direction is it from the house? The answers to these questions will help you determine what landscaping alterations you can make to help conserve energy. Details on how to assess these factors are on pages 46-49.

In-Between

■ **Doors and windows:** Between the inside and the outside of your house are the doors and windows. They are a major source of heat loss, around their edges and through their glass. If your house is typical of most American homes there's little or no filler for the cracks around these openings. All these big, heat-devouring holes in your walls should be sealed along the top, bottom, and sides so that, when closed, they fit tightly against their jambs. Weatherstripping is the name of the game, here, and all the how-to's are discussed on pages 24-27. If, when checking your doors and windows, you find existing weatherstripping, note whether it is in good repair or needs to be replaced.

Now look at the glass on your windows and all doors to the outside. Are they single pane or double pane? Are there storm windows or doors? As you draw each window onto the map of your house, note what kind of window it is; whether it faces north, south, east, or west (you can approximate, if need be); how many windows face in each direction; and what amendments currently exist around the window—awnings, overhangs, curtains, shades, shutters, and so forth. Later on, you'll learn how to turn any such amendment into an energy-conserving advantage (see pages 32-37).

Other things to look out for are: Which windows admit the most light? Which admit the least light? Which let in the sun? Which don't? Which have views you want to keep? Which don't? This information will help you decide what changes to make at your windows.

Inside

■ **Fireplace and chimney:** A fireplace and chimney may seem romantic, but what they really amount to is a large hole in your ceiling, through which a huge amount of heat escapes. A fireplace loses about as much heat as it produces if you use it often; and if you use it only now and then, it loses even more heat. From a conservation standpoint, your best bet is to just plug it up.

However, if you don't want to do that—or if you use your fireplace enough to warrant keeping it unplugged—there are a few things you can do to keep the heat indoors.

First see if you have a damper, one that fits snugly into the flue. What kind of fireplace screen do you have—wire mesh, glass? Is your fireplace vented to the outside? Check for ducts in the floor or on the wall. Do they work? Is your chimney clean? Chapter 6 outlines the ways you can maintain your fireplace and make it as energy-efficient as possible.

■ **Heating system:** Now turn on your heating system and check each outlet for heat. Are there any that don't seem to be working? If you have forced-air heating and some room is not getting warm air through its duct, the answer to your problem may be in the furnace or in the venting system itself. Are the heating vents free of dust and dirt? Make a note to clean them if they're not. Is your thermostat working properly? Note if you've had any problems with it—if you have, it should be checked by your heating service person.

■ **Other heat sources:** As you walk through your living space, draw on your map stoves, heaters, chimneys, and recessed heating or lighting fixtures, ceiling fans, in-wall heaters, and other sources of heat. This drawing will come in handy when you get to the attic.

■ **Water heater:** Find your water heater and put it on your map. Now write down the temperature at which you now have it set, and indicate whether or not it is insulated. (If your water heater is located in the basement, wait until you get down there.) Note also whether any of the pipes are insulated, and if the insulation is well-sealed.

■ **Waterbed:** If you have a waterbed, draw it on the map and write down its temperature, too. A waterbed costs about $5 a month to heat. However, you can keep the expense of keeping those hundreds of gallons of water warm by letting the sun shine on it or by covering it when the sun is not shining.

Attic

■ **Proper clothing:** Dress properly to audit your attic. Not black tie and tails—sturdy clothing that will protect you and that you don't mind getting dirty. You will need:

1. A hard hat. Long nails protruding from unfinished roof beams can be not only painful but downright dangerous. If you can't get a hard hat, at least wear a loose hat that will press in on your head and let you know when you are about to bump into something.

2. Eye goggles. Wear these to protect your eyes from protruding objects and to keep out dust and other foreign particles.

3. A respirator. If your attic already

Cold air can enter when cracks develop where two building materials meet.

Look for infiltration wherever ducts or other supply systems pass through the walls.

P G & E conservation analyst Noelle Buer measures a window for weatherstripping.

has some insultation, a respirator is especially for you. Tiny particles of insulation can be extremely irritating to the lungs. Even if there's no insulation up there yet, if you haven't been in your attic for a while you're going to kick up a lot of dust, and the respirator may save you from a prolonged bout of sneezing. If you don't have a respirator, cover your mouth and nose with a handkerchief.

4. Gloves. The same particles of insulation that can irritate your lungs can also irritate your skin. Besides, you're likely to disturb some spiders and insects, and the wood will probably be splintery. The best type is the sturdy construction gloves available for a few dollars at any hardware store.

5. Long-legged pants and long-sleeved shirt.

6. Sturdy shoes. Nails can stick up from floors as well as down from roofs.

■ **Proper equipment:** You'll also need to bring along the following equipment:

1. A ladder. A good sturdy ladder is useful for reaching high ceilings, vents, and so forth.

2. A flashlight. Make sure it has batteries that work.

3. A strong board. Use this to straddle the narrow joists in the floor, if your attic is unfinished. (Or you can forego this and just stand on two joists at a time, instead.)

4. A tape measure, pencil and your map, of course. Measure your findings and note them on the map.

■ **Joists:** Some attics are entered through a door from a room on the topmost floor. But with most attics, you climb an additional flight of stairs, and then climb a ladder to a door or hatch. If this is the case with your attic, be sure to use your flashlight. Beam it in even before you pull your whole body up there. Now, while still perching on the ladder, take a look around. Is there some sort of support to stand on? In most unfinished attics, the "floor" is nothing but joists running above the plasterboard that forms the ceilings of the rooms below. Don't step on this plasterboard—it won't hold your weight. And don't stand on just a single joist—it might not hold your weight, either; or it might flex, popping nails on your ceiling boards and causing cracks below. In any case, you might lose your balance. Instead, you can stand on two joists at a time, or else bring along a strong board that's wide enough to straddle a couple of joists, and walk on that.

Now bring yourself fully into the attic and take a look at the joists. How far apart are they? In most houses, they will be 16 inches or 24 inches apart. If yours are *more* than 24 inches apart (which is likely in some older houses), check to make sure that the ceiling below will stand up to the weight of new insulation.

■ **Roof:** Now look at the inside of the roof. Note every single hole. A good way to recognize holes is that you can see daylight through them. Also look for leaks or any other sort of damage in the roof. Check around the flashing and the ventwork, as well. *All* roof repairs must be made before you even think of adding insulation. This is especially necessary if water has already gotten into the house, because the water will ruin your insulation as well as the structure of your house.

Be on the lookout for any signs of dry rot along the beams, where water may have gotten into the attic. Dry rot looks like a white, powdery accumulation on top of spongy, punky, softening wood.

Is there any existing insulation? If so, what kind is it? How deep? (For instructions on recognizing and measuring different kinds of insulation, see pages 38 and 42.)

Make sure all recessed heating or lighting fixtures are clear. You noted them on your map when you were in the living space, and from here in the attic you should be able to spot their locations again.

When you have noted all the problems with the roof, there are other things to look for. There's no particular order; just be sure you get them all.

■ **Vents:** Make sure all air vents are clear and open. Make note of where the vents are so that you don't cover them with insulation later on. The purpose of the vents in the attic is to keep the air fresh, and to keep moisture from building up (otherwise it might rot your insulation and housing structure). And in hot weather, vents keep your attic from becoming a hotbox that filters unwanted heat to your living quarters.

In fact, vents are so important that if you don't have any, add no attic insulation until you've put some vents in. Although they need not conform absolutely to federal energy standards, you should have approximately one square foot of vent for every three hundred square feet of ceiling area. There's really no such thing as "too much" venting in a properly insulated attic, but there is such a thing as "too little." (For more details on vents, see pages 58-59.)

■ **Wiring:** Examine your wiring. If it's in good condition, you can lay insulation right under it. But if it's worn, call an electrician to tend to it before you lay any insulation. In fact, any time there's

When doors are not sealed tight in their frames, infiltration occurs.

Doors and windows that do not close completely cost you energy dollars.

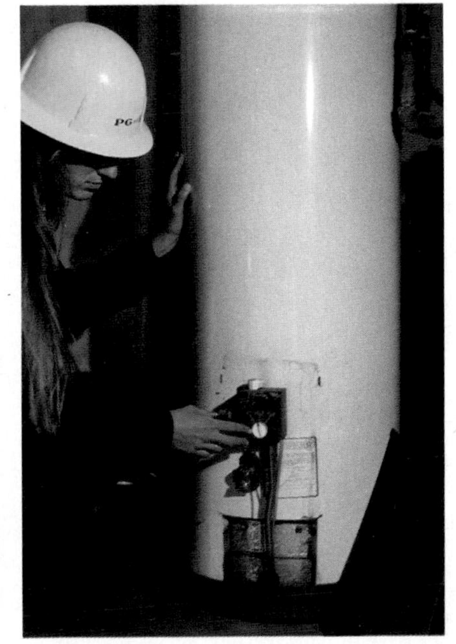
Most domestic hot water needs only require a low to medium temperature (about 120°).

plumbing or electrical work to do, it's wise to enlist some expert guidance. Electricity is dangerous if you don't know exactly what you're about, and since both electrical and plumbing repairs can involve the structure of your house, errors or misjudgments can be quite costly.

If your house is more than 25 years old, you may have knob and tube wiring. In this system, tubes containing the wires run through or across your joists, and you can see glazed ceramic conductors on top of them. If you hire a contractor to insulate your attic, be sure to point out the knobs and tubes. If you plan to insulate your attic yourself, when it comes time to actually lay down the insulation (which most likely will be fiberglass batts), you'll need to stop at every tube and cut the batts to fit down next to the wiring. You'll also need to keep the aluminum vapor barriers away from the wiring, so that you don't short-circuit your whole house.

■ **Heat sources:** When you went through your living space, you made note of all stoves, heaters, recessed lighting fixtures, and other sources of heat. This was so you'd know their locations well enough that you wouldn't put the insulation too near them. Most insulation materials are flammable to one degree or another. Fiberglass isn't flammable, but it pays to keep even this material away from heat sources. It never hurts to be too careful about safety.

So when you insulate your attic, you'll want to install barriers around these heat-producing fixtures—especially if you're using blow-in cellulose. This will keep loose fragments of the insulating material from falling into the fixtures, which would create a very real fire hazard. By noting all fixtures, vents, wiring, etc., you'll know how many barriers you'll have to build. You'll find further information on insulation on pages 38-45.

■ **Vermin:** Look around for signs of vermin. Like it or not, your house *could* be the happy home of a few families of rodents or insects. And you might as well evict them before you supply them with all that nice, warm, nest-like insulating material.

■ **Holes:** Finally, look around in your attic for any holes that lead to living areas. These should be plugged to reduce infiltration, no matter what type of insulation you use, but they especially should be plugged if your insulation will be loose fill—otherwise the insulation will sift right through the holes.

Okay—come on down from the attic, now. (Remember to put the hatch back on.) Take off your hard hat, your goggles, your respirator, and any clothes that have gotten really dirty. Have a cup of coffee and make sure your notes are legible.

Basement

■ **Furnace:** The furnace is the most important part of any basement audit. (Some furnaces are in first-floor closets instead.) Take a look at the filter. Is it clean? Look at the ductwork, especially at the joints. Is your duct tape secure, or is it damaged or missing? Are there any leaks? If you aren't getting enough heat through one vent upstairs, it could be because some ducting is missing or broken.

You might consider having a service person check your furnace for efficiency—either now, as part of your energy audit, or later, after you've read through pages 52-55. If you have a gas furnace, ask the service person to show you how to turn your pilot light on and off. Turning the pilot light off completely in months when you won't be using the furnace can save you energy as well money.

■ **Hot-water heater:** Check your hot-water heater. Is it insulated? If its pipes run through an unheated space and there is any chance that the space may freeze, the pipes should be insulated as well. As a side benefit, that bit of insulation will reduce your waiting time for hot water when you turn on the tap.

■ **Ceiling:** Check for insulation on the basement ceiling above you. If the basement is unfinished, consider it outside space, just like your attic or garage. You should insulate your living quarters against the cold that can seep up in winter.

■ **Vent holes:** Finally, examine the holes where the heating vents go through to the floor above you. Sometimes they are installed so sloppily that air comes in around the vent grille. If you find this problem, note on your map that rifts here should be caulked, or stuffed with pieces of insulation.

Okay, come upstairs again. You've just completed the first step of your own, personal, walk-through home energy audit. Now that you know what shape your house is in, let's see what that means in terms of energy and dollar costs and savings.

Take a look around even before you pull yourself completely up into the attic.

How far apart are your joists? In most American houses they are 16" or 24" apart.

Measure the joists on the basement ceiling as you measured those on the attic floor.

The Financial Picture

Obviously, there is no such thing as an "average" American house. Therefore, all the following figures should be taken as guidelines only.

In the "average" American frame house, with no existing insulation, 30 to 35 percent of all heat loss occurs through the ceiling; 25 to 30 percent occurs through the floors; 15 to 20 percent through the walls; 10 to 15 percent through infiltration; 10 to 15 percent through doors and windows; and about 2 percent through the ductwork.

Unless otherwise noted, the costs for energy conservation improvements listed here are for materials only. We are assuming that you will do the work yourself. If you hire union labor for any of these jobs, expect to pay about $40 per hour per person for that help. If you hire a semiskilled handyperson, expect to pay about $10 per hour. If inflation continues to climb, all costs—materials and labor—are likely to do the same. Nor is there any certainty about the number of hours your employee will require to accomplish a given task. How long the job will take depends on the employee's skill, partly on the nature of the job and the job site, and partly on the nature of the employer—you.

Costs will vary according to locale and the nature of your heating system. In general, costs are as follows:

Attic insulation (fiberglass batts): 18¢/square foot to R-11; 25¢/square foot to R-19. (For an explanation of R-values, see page 38.)

Floor insulation: Same as attic insulation.

Wall insulation: Same as attic insulation, assuming: 2 by 4 studs; open wall; and bare studding. These are unusual conditions, however. More often, the wall will be sealed and you will have to hire a contractor to blow in loose fill insulation or foam-in-place insulation (see page 39). For contractor-installed, blown-in cellulose, figure 70¢/square foot. Foam-in-place (depending on the foam, the contractor, and your house) will cost more.

Plastic storm windows, including screen-type framing (see page 32)**:** 50¢-foot, including frame.

Storm doors: $60 per door and up.

Weatherstripping (see page 22)**:** Flexible—$3 to $4 per door; rigid-$10 per door. The higher cost of rigid weatherstripping comes with good news and bad news. The bad news is that rigid stripping is harder and more complicated to install than flexible stripping. The good news is that it lasts far, far longer than flexible stripping, and will easily earn back its cost.

Water heater insulation (see page 63)**:** $20 per heater.

Clock (timer) thermostat (see page 65)**:** $20 to $60, plus installation.

Ductwork insulation: 50¢-60¢ per linear foot without air conditioning; $1 per foot with air conditioning.

Caulking: $1 to $5 per tube.

In our "average" house, the costs and savings for these energy-conserving amendments are as follows.

But your house isn't average; most likely, you have a unique combination of circumstances that you'd like to correct or improve upon in order to save energy and money in the coming years.

Because every house and every person's use of energy is different from all others, there is no way we can tell you *exactly* what improvements to make, or *exactly* what they will cost, or *exactly* how much energy and money such improvements will save you over how long a period of time. But we can give you a *general* formula that will enable you to make your own, fairly accurate, assessment.

1. Add the totals of all your utility bills (both gas and electric) from the past 12 months. If you don't have a checkbook record of the amounts, you can find out what these are by calling your utility company or companies. In your total, include any money you spent on back-up or auxiliary heating, such as wood, coal, or propane, but don't include the price of stoves, fireplace screens, or other tools or machines.

2. Select one month when you did no heating and no cooling through any of your systems, but not a month when you were away on vacation or had all your relatives visiting. Pick an "average" month without heating or cooling. If you have no cooling system, pick a summer month; if you do have a cooling system, choose a spring or fall month.

3. Multiply that month's utility bill(s) by 12 (for 12 months).

4. Subtract the result you got in Step 3 from your total in Step 1. The number you are left with in Step 4 is approximately what your heating and cooling costs were in the past year.

5. To find out approximately what *dollar* savings you can realize per year from any of the amendments we mentioned earlier, simply multiply your answer from Step 4 (your heating and cooling bill for the past year) by the *savings percentage* indicated for that amendment. Don't forget the decimal point.

For instance, if the number you come up with in Step 4 is $1000.00, and you want to know how much money you might save per year by adding attic insulation in Atlanta, multiply $1000.00 by .43. The answer is $430.00.

If you want to know how long it will take for your conservation improvement to pay back the money you have invested in it, simply divide your answer from the preceding paragraph ($430.00) into the multiple ($1000.00). In this case, your answer is about 2⅓ years, assuming your fuel costs remain stable. If your fuel costs rise, you will save more money on your conservation investment, and it will pay you back faster. For example, if your fuel costs double the day after you install attic insulation in the problem above, your savings will be closer to $860.00 than to $430.00, and your conservation investment will pay back in about 1¼ years.

Remember, these numbers are guidelines for an average house in each of the specific areas mentioned in the chart. The actual numbers you come up with, and the actual savings you achieve, may be greater or smaller than our numbers.

There is also an interactive effect, which will alter your own final, real numbers. That means that when you insulate both your ceiling and your walls, for instance, the *combined* savings will not equal the savings indicated for insulating the two areas separately. In other words, if you can save $100.00 per year by insulating the ceiling, your heating costs will *already be lower* when you start to insulate your walls, and the savings you will realize from insulating your walls will be a little bit less than the original figure projected.

Weatherizing Your Home: Savings per Year

Because the costs of materials, labor, and fuels vary from one region of the country to another; and because every house is unique, the figures below reflect general national averages for weatherizing a 1400 square foot ranch style home in 1980. Our model house begins with no insulation and is improved to R-19 ceilings and R-11 walls and floors. Except where noted costs are for materials only. All numbers are approximations and are to be used only as guidelines.

Improvement	Cost	Dollar Savings/Year	Percent Fuel Bill Savings/Year
Attic (Ceiling) Insulation	$300-$400	$100-$350	10%-20%
Floor Insulation	$300-$450	$60-$110	3%-5%
Wall Insulation (Contractor Installed)	$1000-$1500	$85-$150	3%-6%
Storm Windows & Doors	$150-$600	$50-$300	5%-10%
Weatherstripping	$20-$80	$10-$50	1%-2%
Water Heater Blanket	$20-$25	$5-$20	1%
Clock (Timer) Thermostat (10° Nighttime Setback)	$30-$65	$30-$150	1%-5%
Ductwork Insulation	$20-$50	$10-$25	1%
Caulking	$20-$100	$10-$25	1%
Totals	$1860-$3270	$360-$1180	26%-51%

A Simple Check-list for Your Walk-through Home Energy Audit

So that you won't have to be flipping through the pages of this book while you stand on a ladder, probing your attic ceiling for dry rot or looking up your chimney, we offer this check-list. It shows at a glance what to look for, and gives you space to make notes on your findings. Every item contained in this check-list is explained in the walk-through audit.

Outside

This includes the outside of the house, and the area around it.

Cracks and Joints: In particular, look for cracks and joints that need to be sealed with caulking compound. If caulking already exists, note whether it is in *good* repair (complete and without cracks), *fair* (mostly good but with some cracks or missing pieces), or *poor* (missing or badly broken).

- Are there cracks in the outer walls? Yes______ No______ ■ Where? ______

	Good	Fair	Poor
■ What is the caulking like where two or more different kinds of building materials come together?	___	___	___
■ Where concrete foundation joins wooden, aluminum, or other siding?	___	___	___
■ Where window or door frames fit into walls?	___	___	___
■ Which ones? ______	___	___	___
______	___	___	___
______	___	___	___
■ Where roof joins siding?	___	___	___
■ Where? ______	___	___	___
______	___	___	___
______	___	___	___
■ Where siding meets chimney?	___	___	___
■ Where holes in roof or walls accommodate wiring, ducts, pipes, tubes, or hoses?	___	___	___
■ Where? ______	___	___	___
______	___	___	___
______	___	___	___

Landscaping:

- Are you on a hill or slope? Yes_____ No_____ ■ What direction does your house face? North____ South____ East____ West____
- What are the directions of the prevailing winds? Winter______ Summer______
- Are there trees on your property? Yes______ No______ ■ For each tree, or stand of trees, note the following:
  What kind of tree is it?______ How tall is it?______
  How far from the house and in what direction?______
  Is it mature or immature?______ Is it deciduous or evergreen?______
- Is there shrubbery next to the house? Yes______ No______ ■ Are there vines next to or on the house? Yes______ No______
- Are there fences on your property? Yes______ No______ ■ How high are they and where are they located?______
- Do you have water on your property, such as a pond? Yes______ No______ ■ Where is it located? ______

Swimming Pool:

- Do you have a swimming pool? Yes______ No______ ■ Is it heated? Yes______ No______ By what method? ______
- Do you have a swimming pool cover? Yes______ No______ ■ What type?______
- Is it in good repair? Yes______ No______

Living Space

This includes all parts of the house in which people actually live and work. It does *not* include unheated garages, attics, basements, or crawlspaces.

Windows and Doors:

- Are your doors and windows weatherstripped? Yes______ No______ ■ What condition is the stripping in?
  Good_____ Fair_____ Poor______ Which doors and windows are not weatherstripped?______
  ______
- Do you have single-glazed windows? Yes______ No______ ■ Where and facing in which direction? ______
- Double-glazed windows? Yes______ No______ Where and facing in which direction? ______

■ Storm windows or doors? Yes_____ No_____ ■ What type are they? _____

■ Where and facing in which direction? _____

■ Awnings and/or overhangs? Yes_____ No_____ ■ Where and facing in which direction? _____

■ Curtains, shades, shutters, etc.? Yes_____ No_____ ■ For each window, what type of covering do you have?_____

_____

Fireplaces:

Yes_____ No_____ ■ Size (s) _____ ■ Damper? Yes_____ No_____ ■ Glass Screen? Yes_____ No_____

■ Is fireplace vented to the outside? Yes_____ No_____ ■ Is the chimney clean? Yes_____ No_____

Heating System: (Check any that apply.)

Furnace_____ Gas_____ Forced air_____ In-wall_____ Electric_____ Steam (radiator)_____ Radiant_____

Oil_____ Hot water (baseboard)_____ Other_____ (Propane, coal, wood, solar)

■ Are heating vents clean? Yes_____ No_____ Is thermostat working properly? Yes_____ No_____

Water Heating System:

■ Gas_____ Electric_____ Solar_____ ■ Temperature setting? _____ ■ Insulated? Yes_____ No_____

■ Pipes insulated? Yes_____ No_____

Attic

An upper crawlspace may be considered an attic, if that's what your house has. By "attic," we mean the topmost, probably unfinished, portion of your house.

■ Is the attic "floor" insulated already? Yes_____ No_____ ■ What kind of insulation? _____ ■ How deep?_____

■ Can you see any holes between the attic and your living space? Yes_____ No_____ ■ Where? _____

What size? _____

■ Can you see the joists? Yes_____ No_____ ■ How far apart are they? 16"_____ 24"_____ More than 24"_____

■ Can you see the wiring? Yes_____ No_____ ■ What condition is it in? Good_____ Fair_____ Poor_____

■ Do you have knob and tube wiring? Yes_____ No_____ ■ Can you find any holes in the roof? Yes_____ No_____

■ Where? _____

■ Do you have any vents? Yes_____ No_____ How many? _____ Where?_____

What size? _____

_____

■ Can you see any dry rot? Yes_____ No_____ Where? _____

■ Are there any recessed lighting or heating fixtures that might heat up and endanger insulation? Yes_____ No_____

■ What and where? _____

■ Can you see any evidence of vermin? Yes_____ No_____ ■ What and where? _____

Basement

A lower crawlspace may be considered a basement, if that's what your house has. We use the term to designate the lowest, probably unfinished, portion of your house. If your heating plant or duct work is located somewhere other than the basement, you should still answer the questions here that pertain to it.

Furnace: ■ Is the filter clean?_____ or dirty?_____ ■ Ductwork: Leaks? Yes_____ No_____ Where? _____

■ Duct broken or missing? Yes_____ No_____ Where? _____

■ Tape damaged or missing? Yes_____ No_____ Where? _____

■ When was the unit last serviced? ■ Less than one year_____ ■ More than one year_____ ■ Don't know_____

Hot-Water Heater:

■ Insulated? Yes_____ No_____ ■ Are the pipes insulated? Yes_____ No_____

Basement Ceiling:

■ Insulated? Yes_____ No_____ ■ What kind of insulation? _____ ■ How deep?_____

■ Are the holes around heating vents filled in? Yes_____ No_____ ■ With what? _____

■ Is the filling good_____fair_____or poor_____?

Your house is your protection against the elements. When winter's cold and summer's heat leak in, they steal your comfort along with your energy dollars. It takes just a little investment and know-how to plug those leaks.

Plugging the Energy Leaks: The House

As you may have discovered when you did your energy audit, your house is an integral system in its own right. Just as one problem can lead to another (for example, a leaky roof may lead to rotting walls), so one solution encourages another (to continue the example, fixing that leak in your roof will not only help prevent your walls from rotting, but it will also keep the insulation from spoiling).

Maintaining your house in good repair will make your life more comfortable and also save you energy and money, both now and in the future. After all, a solid house is a far better investment than a rickety one. And if you undertake some of the maintenance and repair work yourself instead of hiring it out, you may find that your house becomes a new source of pride and satisfaction.

Your house protects you from the elements—the roof and walls keep cold, heat, rain, wind, dust, and so forth at bay. However, you wouldn't want your house to seal you off from the outside world entirely, so there are windows to admit air, sunlight, and views; doors to let you pass in and out of your protective shell; and chimneys and vents to encourage air circulation and help maintain proper temperatures.

When maintained correctly, all these openings in houses are useful and advantageous. The problems crop up when undesirable holes appear in the shell, allowing the walls to be penetrated by these very elements you intended to keep out.

In this section, you will learn how to deal with the *unwanted* outside air that comes into your house, chiefly in the form of cold air that replaces the warm air you've gone to some expense to heat. Specifically, you'll find out how to seal the energy-robbing holes and gaps that make your protective envelope less efficient than it might be.

Most of the holes in your house may escape your notice because they are so obvious. An ordinary single pane window, for instance, may allow as much as *20 times* more heat to escape than the wall that surrounds it, even though the wall is far larger than the window. An uninsulated roof (or attic floor) may cost you a third of all the heat your house consumes. Poorly-sealed cracks around the foundation, sill, chimney, or siding may cost another 10 percent or more of your heating. And for the energy it wastes, that ¼-inch gap underneath your back door might as well be a 3 by 3 inch hole in the middle of your wall, because that ¼-inch gap extends about 36 inches long. If you have many little holes like these around all four sides of every door and window leading to the great outdoors, that's like having an open hole the size of a picture window smack-dab in the side of your house.

Energy that goes to heat your loosely-sealed house is truly wasted in the worst way, and the vast majority of American homes are inefficient energy-users for just this reason. It takes little effort and a very small investment to plug those leaks and stop your energy and energy dollars from pouring out through your windows, sliding out under your doors, and spilling through the tiny chinks in the joints of your house. It takes only about one day to insulate the attic of an average American house; and at today's fuel prices, it will take as few as three to four years in energy savings to make up for the expense of the insulating materials.

If the task of tightening your house for energy conservation looks overwhelming at first, take it one small step at a time. After all, as the ancient Chinese philosopher Lao Tzu observed, "The journey of a thousand miles begins with a single step."

Weatherstripping

Weatherstripping is one of the best things you can do for your house. It involves applying materials to the sides and insides of doors and windows to create tighter seals with their jambs and casings. A good job of weatherstripping will keep unconditioned air out and conditioned air in. But beyond that, there are additional advantages. Because weatherstripping is essentially a seal, it also reduces the infiltration of dirt, dust, noise, and moisture; absorbs some of the shock to glass components when you slam a window or door; helps keep windows and doors from rattling in high winds; and cuts down on drafts.

There are two easy ways to tell whether a particular window or door needs to be weatherstripped: stand in front of it when there's a breeze outside; or simply look at it. If you can feel a draft while standing indoors in front of a closed door or window, you need weatherstripping. If you can see that light is coming in around the door, or that there's no stripping already in place, or that the existing stripping is cracked, damaged, or coming loose, you don't have to wait for a windy day to find out that you have an infiltration problem.

If you plan to do a complete weatherstripping job in your house, don't limit yourself only to the obvious. For best results strip not only those windows and doors that lead outside, but also doors that lead to any unheated space (such as a garage, porch, basement, attic, crawlspace, or unheated—or uncooled—spare room) and unusual "windows" (such as skylights). Keep in mind that weatherstripping does not take the place of storm windows and storm doors. In fact, it's a good idea to weatherstrip them, too.

All weatherstripping materials are available in kits that include hardware; or you can buy the component parts separately. To find the total length of weatherstripping you'll need, measure your window or door frame carefully. If you're buying stripping materials for more than one door or window at a time, make a list of how many strips of each length you will need for every door or window. Then, if you're buying components, simply purchase the total amount and cut it up later. Measure and cut carefully, so that you don't end up with a little too much or a little too little.

Before you begin, make sure you have sufficient weatherstripping material, a hammer, tin snips, a tape measure, and a screwdriver or putty knife. Make sure all the channels are clear of flakes of old paint and other debris.

■ **Types of weatherstripping:** There are many types of weatherstripping, made of many materials and sold in a wide range of prices under numerous brand names. Some kinds of stripping are more durable than others. Sometimes the price reflects the durability, but not always. For example, self-adhering stripping is more expensive than tack-up stripping, but it lets go faster. Some kinds of stripping, such as felt or adhesive foam tape, have to be replaced every year or two. Other kinds, such as tension strips, can last the life of your windows if they are installed correctly to begin with.

In choosing your weatherstripping, make sure you have the right kind for the job you want to accomplish. This chart will help you understand the differences between kinds of stripping. If you still have any doubts or questions after reading this section, ask the salesperson at your hardware store for advice so that you don't come home with exactly what you *didn't* want or need.

Kinds of Weatherstripping for Doors and Windows

Kind	Material	Cost (approximate, per linear foot)*	Installation	Use	Remarks
Felt	Wool or other cloth	5-10¢	Very easy. Glue, nails, or staples.	Anywhere it will not be subjected to severe or frequent friction (such as channels of window sashes). Best around doors and at tops and bottoms of double-hung windows.	Should not get wet, as it may rot. Should not be dragged against hard surfaces, which may tear it. Essentially a stop-gap measure for temporary use. Will have to be replaced in 1 to 2 years under normal conditions. Cannot be painted.
Reinforced felt	Wool or other cloth and aluminum	10-15¢	Same as felt.	Same as felt.	Same as felt, but stronger.
Foam tape	Neoprene, urethane, or vinyl	15-20¢ 10-15¢ 5-10¢	Very easy. Self-adhesive or glue.	Same as felt.	Similar to felt, but more versatile. Can be removed easily but may also fall off easily.

* approximate cost as of June, 1980.

Kind	Material	Cost (approximate, per linear foot)*	Installation	Use	Remarks
Rigid strip 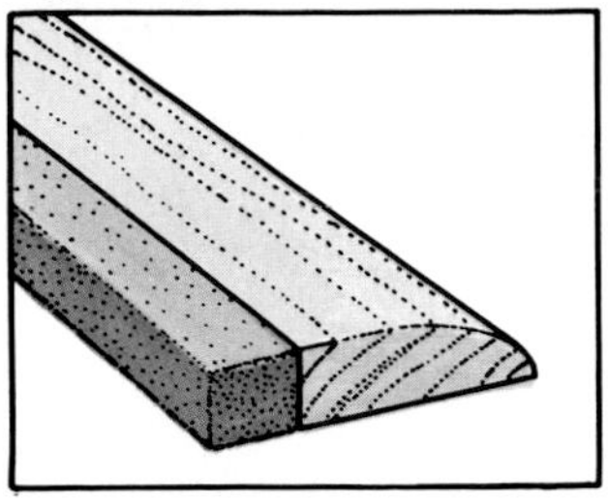	Wood or metal with foam or vinyl edge	20¢ (wood) 35-70¢ (metal)	Easy. Nails or screws.	Doorjambs.	Resilient and wears fairly well. Wood or metal parts should be painted; foam should not be painted.
Tubular gasket 	Hollow, foam-filled, or aluminum-reinforced vinyl or rubber	10-12¢ (hollow) 20-25¢ (foam-filled)	Moderately easy. Self-adhesive or nails.	Anywhere. Best around doorjambs, around sliding windows, where sashes meet on double hung windows. Good at all window junctures.	Quite durable and resilient, withstands friction well; can compress unevenly to fill irregular spaces. Creates good pressure seal.
Metal tension strip 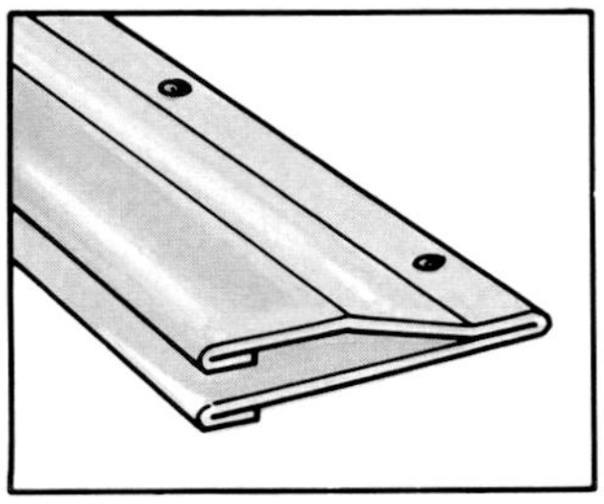	Aluminum, brass, bronze	10-12¢ (aluminum) 20-25¢ (other)	Moderately easy. Nails.	All door and window junctures, especially window sash channels and outside edges of door frames.	Easy to keep out of sight. Extremely durable. Highly efficient. Works best under tension. Does not compress unevenly, and will not seal irregular spaces. After installation, flanges can be pried to create tighter fit—work gently, and do not damage seal or springy quality of strip.
Door sweep 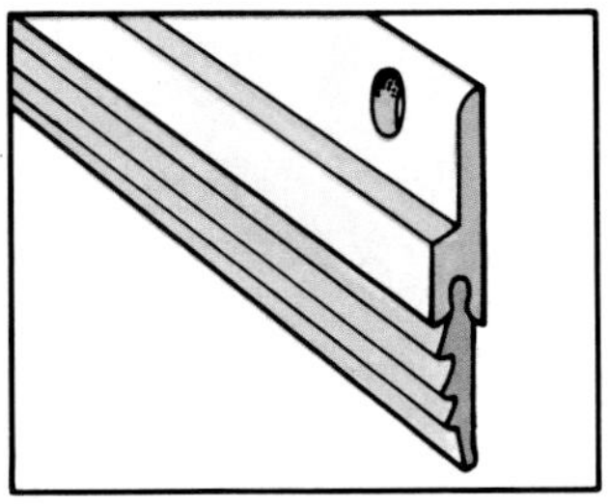	Aluminum reinforced plastic or rubber	60-70¢	Easy. Nails or screws (screws preferred).	Door bottoms.	Efficient, effective, and fairly durable, but inevitably visible and not especially attractive. Must be attached on side *toward* which door opens, and must seal with threshold. Automatic sweeps raise up when door is opened, lower when door is closed, to avoid dragging on carpets, etc., but must not jam too tightly against threshold when door is closed, or mechanism will wear out quickly.
Door shoe 	Aluminum with vinyl insert	$1.50-1.75	Moderately easy. Nails or screws.	Door bottoms.	Very efficient and very durable if installed properly, but door may have to be removed for installation. Requires good threshold for firm seal. Some models have drip caps to shed rain and keep water from running under the door.
Gasket threshold 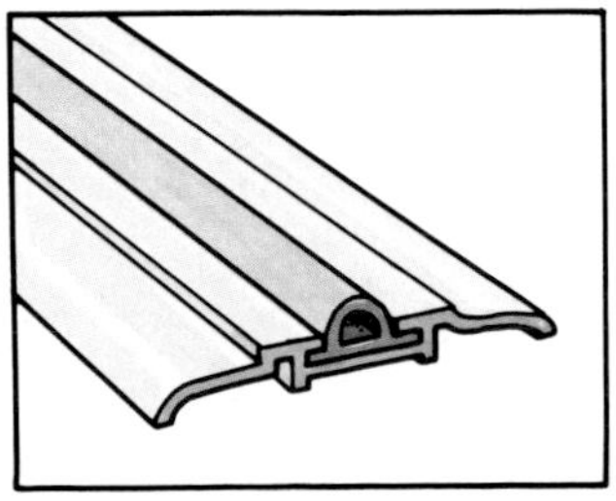	Aluminum with vinyl insert	$2.00-2.50	Moderately difficult. Nails or screws.	Door thresholds.	Gasket similar to that of door shoe, but included in a metal threshold. Very efficient if installed properly, but because vinyl gasket is exposed to foot traffic, less durable than the shoe. As with the shoe, door may have to be removed for installation. Some models have drip caps similar to those on door shoes.

* approximate cost as of June, 1980.

How to Weatherstrip a Window

Every window, like every kind of stripping, has its own peculiarities, so use these instructions as general guidelines, and follow the instructions thoughtfully provided by the manufacturer of the weatherstripping you select.

To illustrate how to weatherstrip a window, we have chosen the double-hung sash and metal tension stripping. This combination allows for the most complete explanation of the whole procedure. As with any weatherstripping, metal strips should be installed with the resilient face pressing against the sash tightly enough to make a good seal but not so tightly that your window will stick.

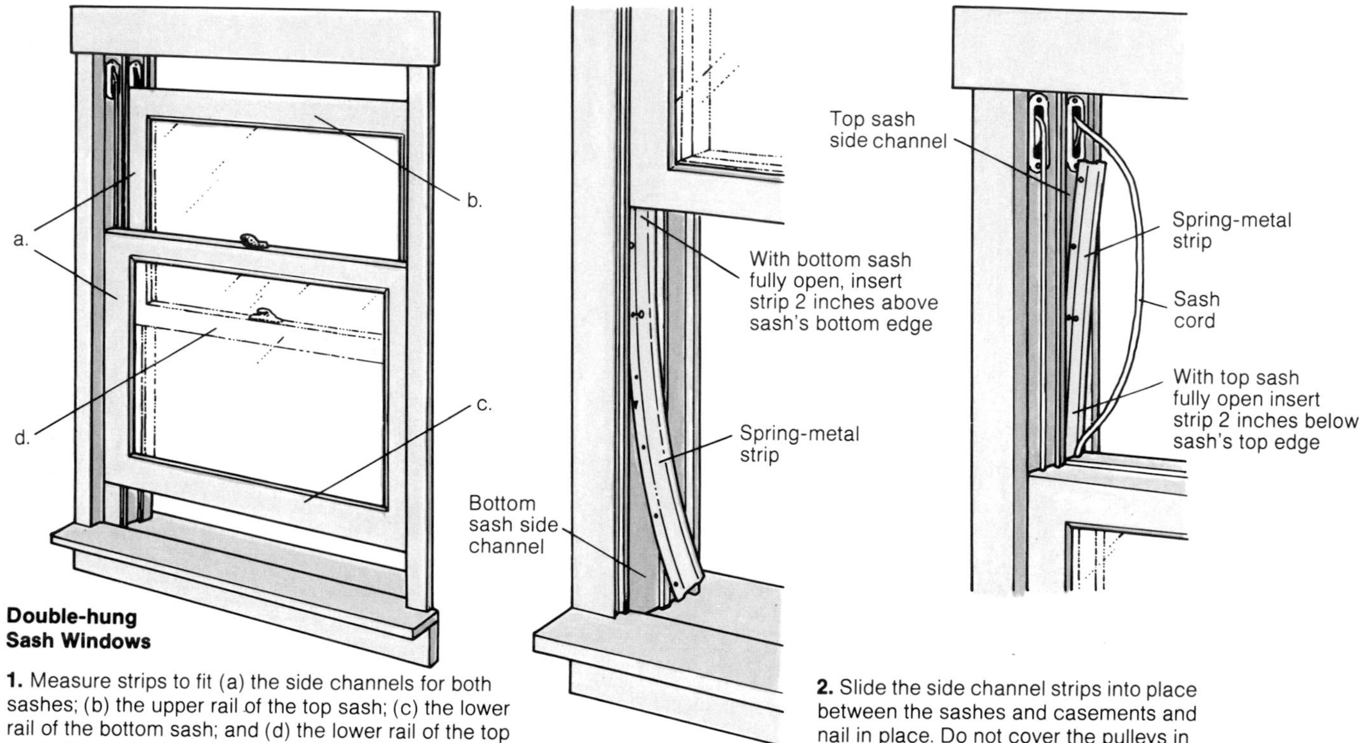

Double-hung Sash Windows

1. Measure strips to fit (a) the side channels for both sashes; (b) the upper rail of the top sash; (c) the lower rail of the bottom sash; and (d) the lower rail of the top sash (the center bar). Cut strips with tin snips.

2. Slide the side channel strips into place between the sashes and casements and nail in place. Do not cover the pulleys in the upper parts of the channels.

3. Slide the upper-rail, top-sash strip into the window's top channel, and the lower-rail, bottom-sash strip into the window's bottom channel, and nail in place. *Or,* as an alternative, nail the upper-rail, top-sash strip to the top of the upper sash, and nail the lower-rail, bottom-sash strip to the underside of the bottom sash.

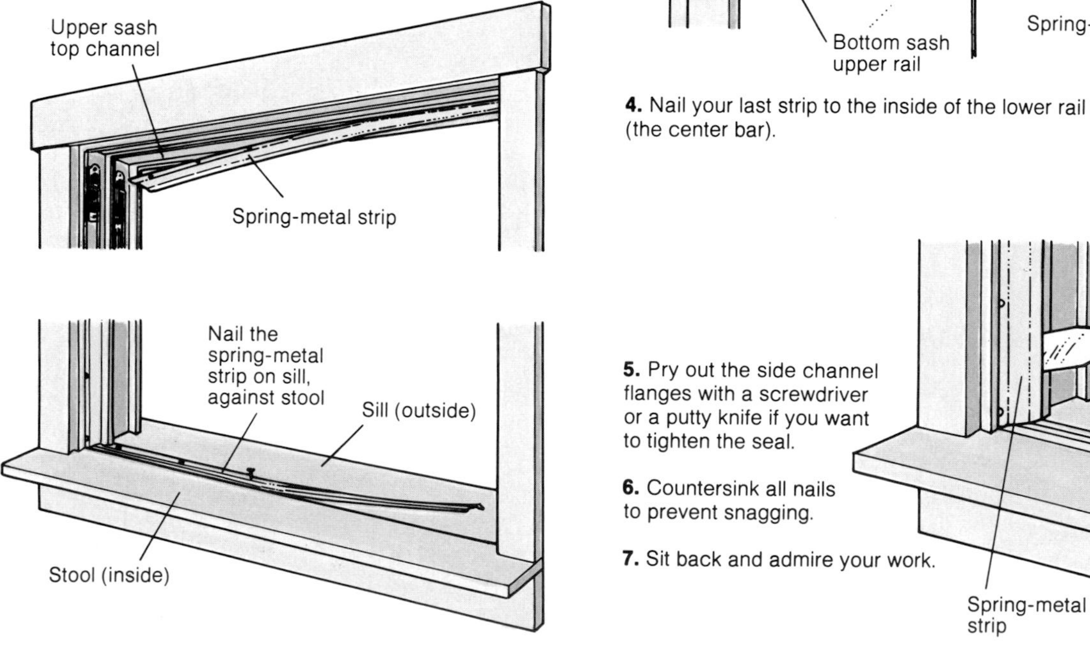

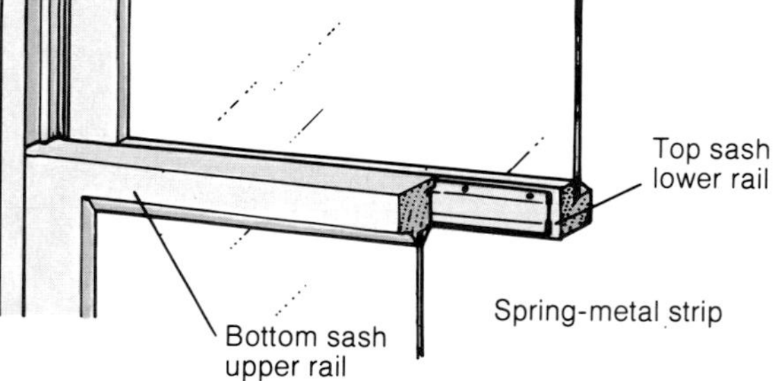

4. Nail your last strip to the inside of the lower rail of the top sash (the center bar).

5. Pry out the side channel flanges with a screwdriver or a putty knife if you want to tighten the seal.

6. Countersink all nails to prevent snagging.

7. Sit back and admire your work.

■ **Sliding windows:** If you want to weatherstrip a sliding window, the steps are the same. Just imagine it's a double-hung sash window that's lying on its side. If only one sash slides, use tension spring stripping in the channel that opens, and seal the three remaining edges of the moveable sash with tubular gasket stripping to create a good seal all around the sliding sash.

■ **Casement windows:** For a casement window or any other kind of tilting window, nail the weatherstripping to the frame with the flange along the edge *toward* which the window opens. If you have a metal casement window and cannot nail into its frame, buy a deeply grooved gasket stripping that can be fitted over all the metal edges of the window frame (available at most hardware stores). To make this stripping hold better, first apply a rubber/metal or a vinyl/metal glue to the frame edge or the gasket channel.

■ **Other weatherstripping methods:** Different types of weatherstripping require varied installation methods. Some of those variations are illustrated below.

Gasket stripping. Vinyl or rubber gasket stripping can simply be tacked all around the sash. To make your gasket stripping less visible, it's best to nail it to the outside of the window frame; but inside or out, it should fit tightly all around, including the lower rail of the top sash. Nail gasket stripping to the window frame with the thick or bulbous side against the sash. Make sure that the rolled edges fit tightly against the window when it's closed. Then add stripping to the lower rail of the top sash (the center bar) on the inside edge, to make a tight seal between the sashes when the whole window is closed. As with tension spring stripping, make sure to strip all edges of the window.

Foam tape. Foam tape or any other adhesive-backed stripping may be simply pressed into place with your fingers. Clean the surface so the tape can adhere. Then apply the stripping, slowly pulling paper or plastic backing off the tape as you go. Do not use this type of weatherstripping where it will encounter friction, such as in side channels—it will wear out quickly, or even pull right off.

Felt stripping. To install felt stripping, you can either staple it in place with an ordinary heavy-duty stapler, or you can nail the felt to the window frame, as with gasket stripping. Also add a length of felt to the inside of the lower rail of the top sash, to block infiltration from between the sashes. However, you should not attach felt stripping to the outside of a window where it is liable to get wet because it may rot, and you would have to replace it sooner than you might otherwise. And as with foam tape, do not use felt stripping where it will encounter friction.

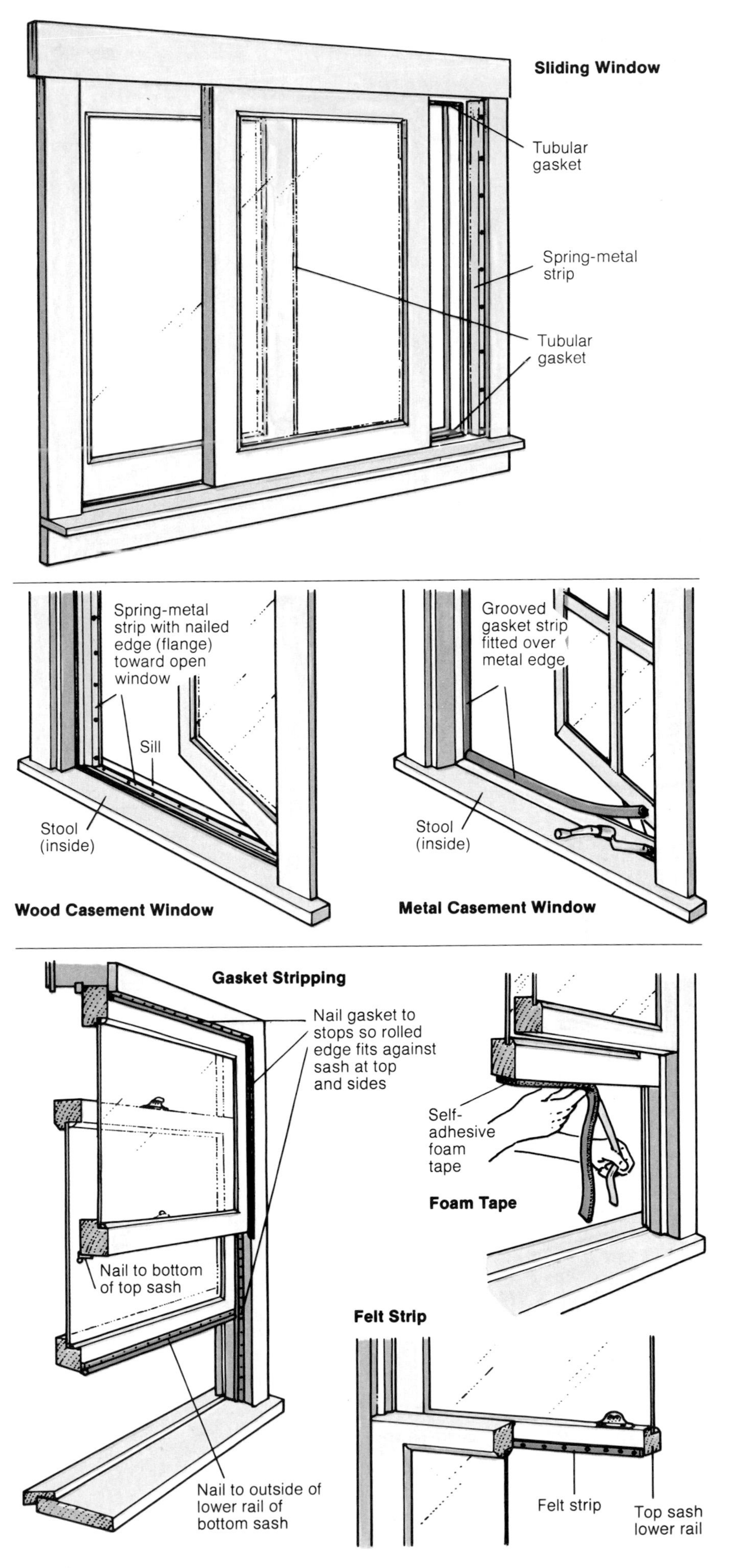

How to Weatherstrip a Door

Although doors tend to be larger than windows, they are generally easier to weatherstrip; unlike windows, these simple rectangles lack double sashes, center bars, and their attendant problems. For most doors, you can follow the same instructions as for weatherstripping a window.

Whether you use rigid insulation with a gasket, adhesive-backed felt, or any stripping in between, first measure the stripping carefully against the frame. Then fit it onto the frame against the doorstop so that it fits tightly when the door is closed, but not so tightly that the door *can't* close. To ensure a proper fit, you may want to first tack the stripping loosely in place and test it by opening and closing the door gently a few times as you work. Then you can drive and set your nails completely.

The exterior door is a major source of infiltration. Its materials don't overlap; its opening is extremely large; and the weight of the door swinging on its hinges over many years tends to loosen screws in the hinges and cause it to sag, making its edges meet the jambs unevenly. For all these reasons, it's definitely worth your while to weatherstrip an exterior door. However, these very same reasons present some unique difficulties.

If your door hangs unevenly, you may have to remove it from its hinges and rehang it, or else the seal between the door and the stripping may not be tight. And if you're going to all that effort, you may want to do it in conjunction with installing a door shoe or a gasket threshold, since to accommodate either of these extremely useful stripping devices, the door bottom must be trimmed —which means that the door must come off its hinges.

1. Begin with the lock strip, the small piece of stripping provided with most kits. Because this piece is the same height as, and placed behind, the strike plate, it will help guide you in measuring and placing your weatherstripping on the side of the jamb where the door is not hinged.

2. When the lock strip is in place, complete the stripping on the sides and top of the doorjamb, keeping the flanges facing toward the door stop. The door bottom will be stripped in another fashion (see below).

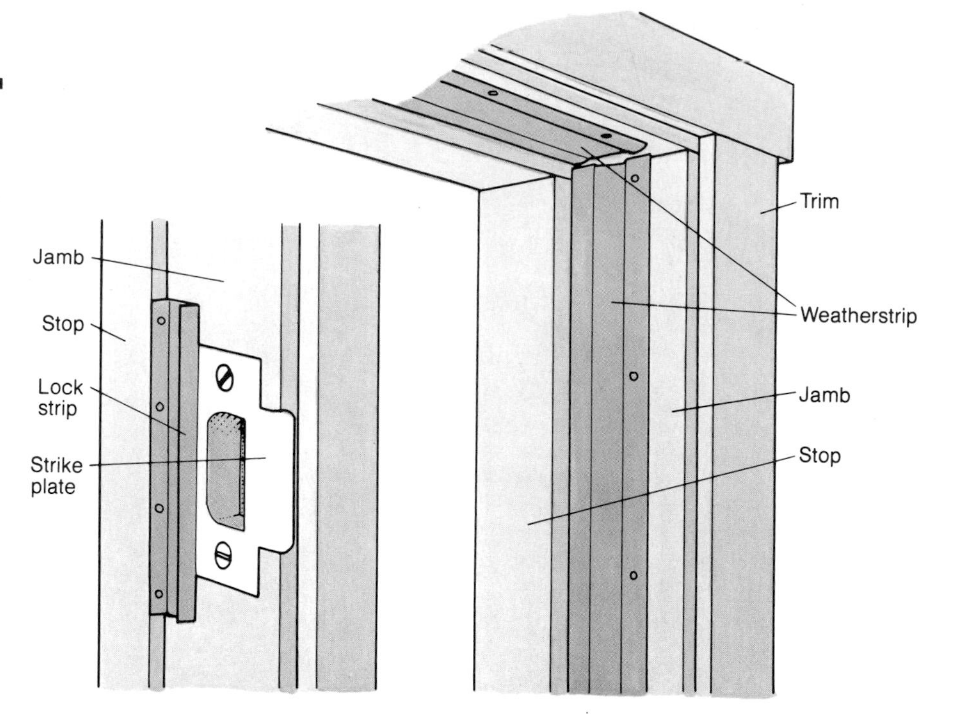

■ **Door sweeps:** Of course, since cold air falls, it is most likely to enter your home through the lowest available cracks and crannies. Virtually any house is subject to this kind of infiltration at the bottoms of doors leading to the outside.

There are various ways to plug these door-bottom cracks, which entail either lowering the bottom of the door, or raising the top of the threshold over which the door passes.

The most common and effective kinds of weatherstripping you can add to the bottoms of your own doors are the *door sweep*, the *door shoe*, and the *gasket threshold.*

The simplest door-bottom weatherstripping, as we've indicated, is a door sweep that you can easily attach while the door is still on its hinges. Just cut the sweep to size and screw it to the base of the door in a position that allows the door to open and close without difficulty.

Adjustable door sweeps are designed to raise up when you open the door inward (over your carpeting, for example), and lower back down when you close the door, creating a tight seal.

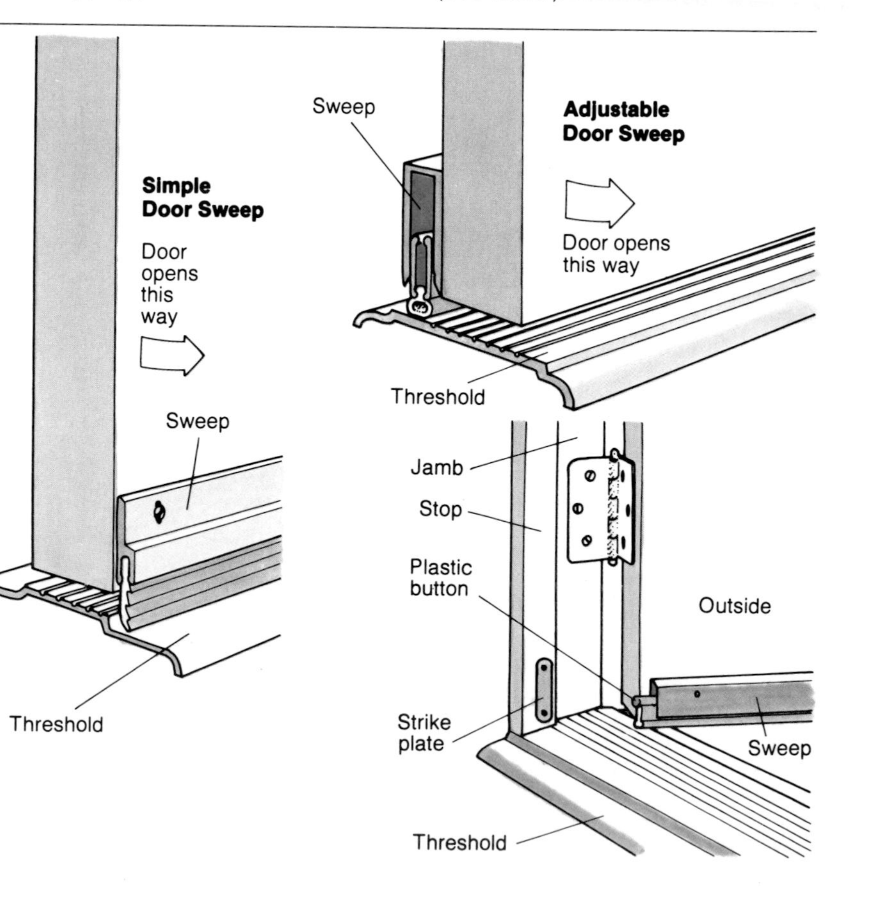

This kind of adjustable sweep is surface mounted to the outside face of your door. To attach it, close your door and measure the width between the stops. Cut the sweep to this size with a hacksaw. Hold the moveable part of the sweep in the "up" position while you cut and be sure to cut the end without the plastic button. Screw the sweep to the door while the door is closed and while holding the sweep in the "down" position so the rubber bottom is pressed snugly against the threshold. The end with the plastic button (which protrudes when the sweep is in the "up" position) must be on the hinge side of the door. Note where the plastic button hits the stop. Open the door and nail the little strike plate (which comes with the sweep) in position at this spot.

■ **Door shoe:** The door shoe is similar to an adjustable door sweep, but it doesn't "sweep." Instead, the vinyl ridges of its tubular gasket make a tight seal with the door sill or threshold.

Attach a door shoe as you would an adjustable door sweep, gauging its fit as you work and trimming it with a hacksaw if necessary. Remove the curved vinyl ridge from the shoe, slide the shoe over the door bottom, screw it on securely, and replace the vinyl so that it makes a tight seal with your threshold.

■ **Vinyl gasket threshold:** If you intend to attach a door shoe, your threshold must be in fairly good condition, and preferably made of wood, so that the ridges of the gasket can grasp and make a tight seal. If you have no threshold, or if yours is worn enough to be replaced, you might consider a gasket threshold, as well as door-bottom weatherstripping. The disadvantage of a gasket threshold is that it gets walked on and eventually wears out.

To attach any threshold, it is often necessary to remove the door from its hinges in order to gain complete access to the threshold area. You may also have to trim the bottom of the door to accommodate the added height of the threshold. If so, first be sure to measure your doorsill, door bottom, and new threshold carefully, so that you know exactly how much of which part you must trim. If you trim too much from your door bottom, you will defeat the purpose of weatherstripping it; if you do not trim enough, your new gasket will wear out very quickly.

If you must trim the door, do it before you trim the threshold. Then cut the threshold to the proper width with a hacksaw, and file it smooth. Center your threshold, and screw it into place.

To achieve a tight seal, you may have to remove the door in any case, and bevel its base about ⅛ inch against the vinyl. *Do not bevel in the wrong direction,* or you may find it impossible to open your door.

Door Shoe

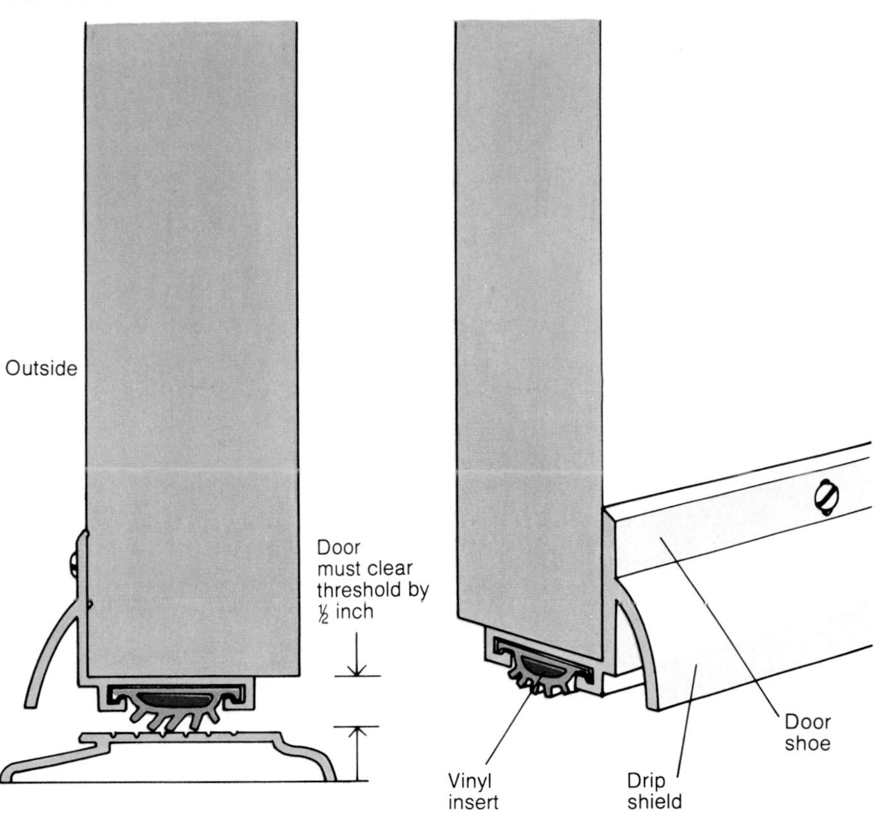

Vinyl Gasket Threshold

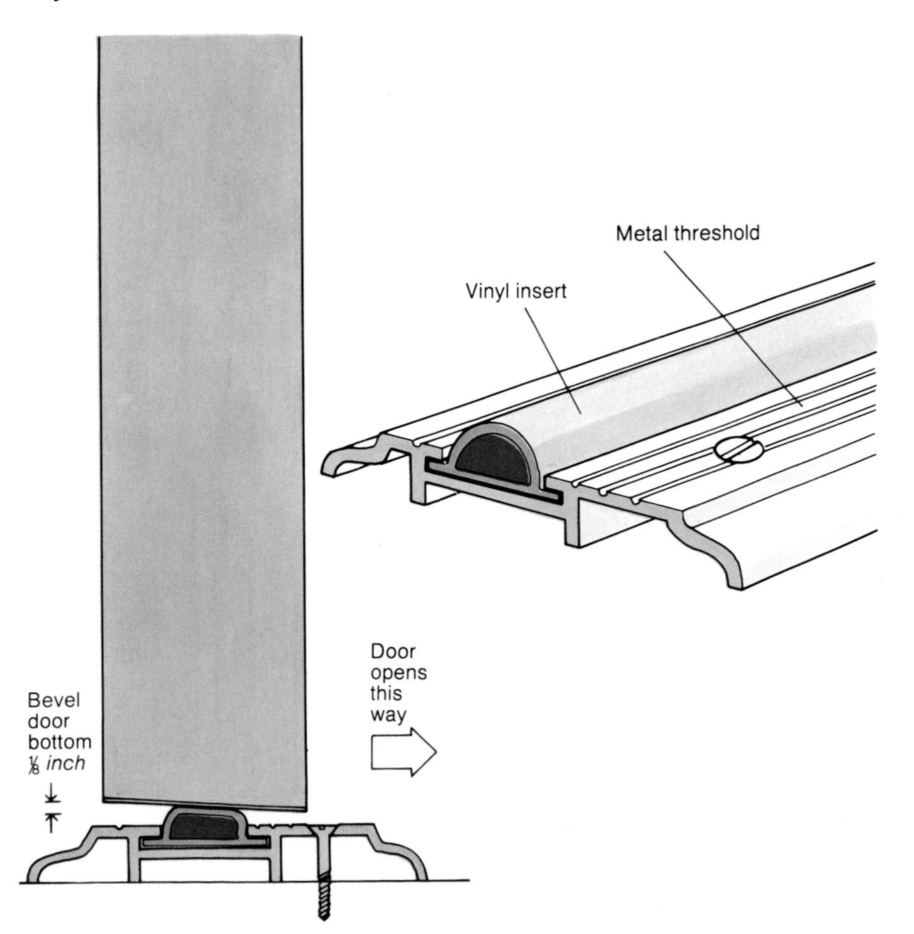

Caulking

Whereas weatherstripping is used to seal cracks between moving parts (doors and windows), caulking is for plugging the constant, unmoveable holes. Like weatherstripping, the purpose of caulking is to make your house more wind-resistant and to help reduce the infiltration of air, dirt, noise, moisture, and insect life.

In the natural course of things, houses settle, ground shifts, wood warps, and temperatures fluctuate. These changes cause materials to swell and contract. Often, the junctures where different building materials meet change size and shape; occasionally they separate permanently. Sometimes a material cracks, chips, or starts to disintegrate. Although these sorts of defects may be minor at the outset, if they are not corrected they can cause far greater difficulties than simple infiltration can. But if you catch them while they are still small, you can fix them all with caulking.

Properly applied, caulking is one of the least expensive ways to preserve the structure of your house. Caulking will protect its frame, sheathing, and even the paint job. And, in combination with weatherstripping, caulking may save your home enough energy within the first year or two of application to pay for all the materials you use.

The best time to caulk is on exactly those days you'd rather be doing something else: the warm-but-not-hot days of late April, May, early June, late September, and early October. Why? These days are sufficiently warm and dry to make the caulk flow, mold, and set easily, but the weather is not *so* hot as to give you sunstroke and make you fall off the ladder.

There are a couple of points to keep in mind for any caulking job. *Buy the best-grade caulk for your purposes*, even if it costs more. Caulk that isn't durable or that fails to accomplish the required task will send you back to do the job again much sooner than you will want to.

Be careful. Sometimes people doing home repairs forget about obvious hazards. When you're standing eight feet up on a ladder, don't lean out too far—always keep your entire torso between the ladder rails.

Do not use caulking near an open flame, and do not smoke while caulking. Most caulking compounds are flammable to some degree, or at least give off noxious odors when in proximity to high heat.

If you must caulk in close quarters, make sure that ventilation is adequate to keep the air fresh. Wash your hands when you have finished working. And always keep these materials out of the reach of children and pets.

■ **Where to caulk:** The illustration below indicates which places around your house to examine for possible caulking. In general, the most likely spots are wherever two different materials come together: where the roof joins the walls; where the walls join the foundation; where any exterior coverings such as masonry and siding meet each other; at corners where separate pieces of the same exterior covering join; where anything—wires, chimneys, pipes, ducts—passes through the building's shell; any place the frame of a window or door meets the siding; any place a porch or other addition joins the main body of the house; at cracks or holes in masonry or siding; around window air conditioners; around air and fan vents; at the roof flashing; around window glass; and any place water might collect, such as the junctures of eaves and downspouts.

Indoors, caulking may be used effectively on the walls, ceilings, or floors, where they have become cracked; where two different materials come together (as where a fireplace meets a wall); and around window and door frames. Even though no energy will be saved by caulking around tubs and

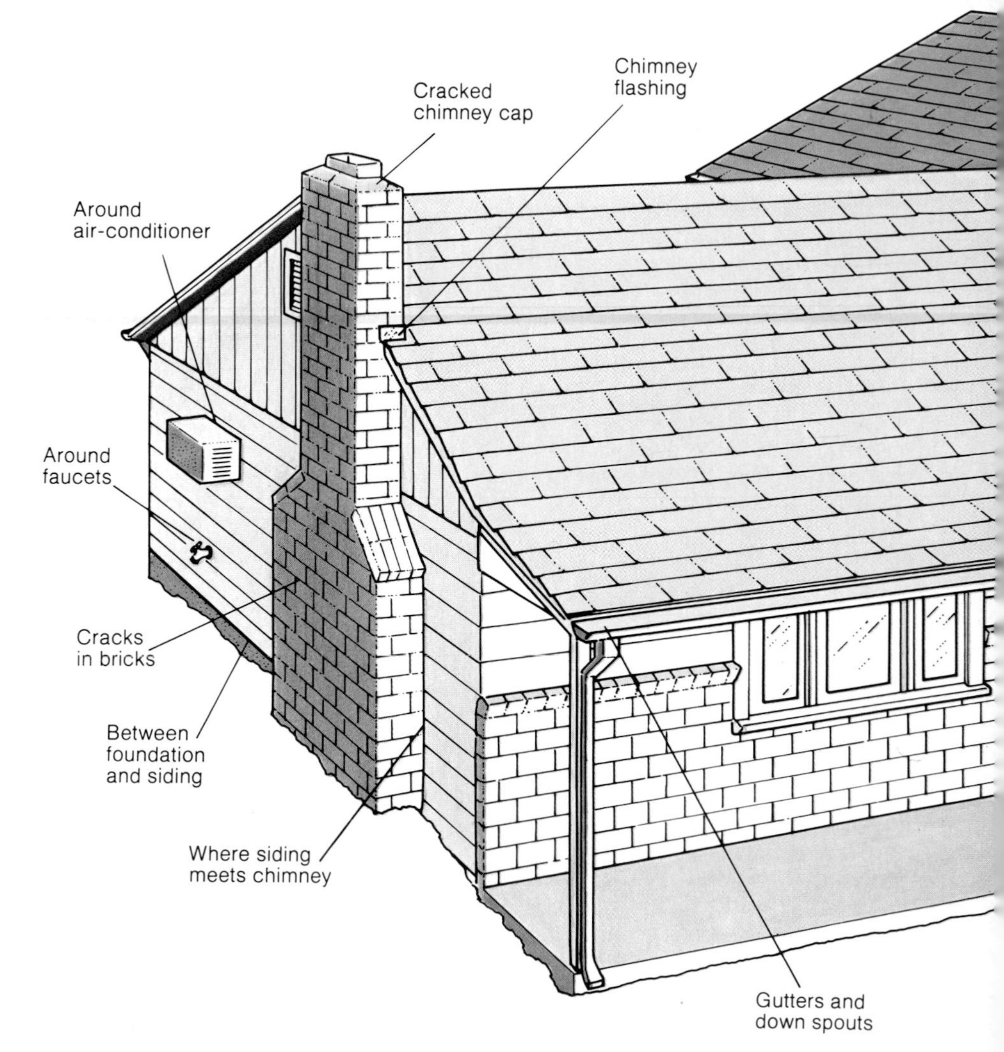

sinks, and where bathroom walls meet ceilings and floors, these operations are worth doing, and are accomplished in the same fashion, and with the same materials, as other applications discussed in this section.

■ Types of caulking: Caulking comes in a variety of forms: the cartridge (which fits into an open-barrel, trigger-operated "gun"); cans; tubes; and coils or rolls. Cartridge caulk, or gun caulk, is generally used for cracks about ⅛ to ¼ inch wide or deep. For narrower or shallower cracks, sealant from a toothpaste-dispenser-type tube is commonly used. Wider or deeper cracks are filled with canned caulking, putty, or other sealant, applied with a flat-bladed knife such as a putty knife. Very wide or deep fissures are sealed with rope or coil caulking. This will work only temporarily, since the rope or coil caulk does not adhere to surfaces; however it can be removed easily when you're ready to make a real repair on that major hole.

Although caulking comes in a variety of forms, by far the cleanest, easiest, and most practical method of application is the caulking gun and cartridge. In fact, aside from some squeeze tubes, an occasional 50-foot roll of oil-based rope caulk (about $4 to $5), and the rare can of oil-based caulk (about $5 a quart), the cartridge is the only form of caulk some leading building suppliers carry, and may be all you can get.

All prices suggested in the chart on page 31 are approximate costs for a single, standard, 11-ounce cartridge of caulk. Except for the butyl rubber, which is quite sticky and therefore a bit more difficult than other caulks to apply, most caulks can be applied with about the same facility.

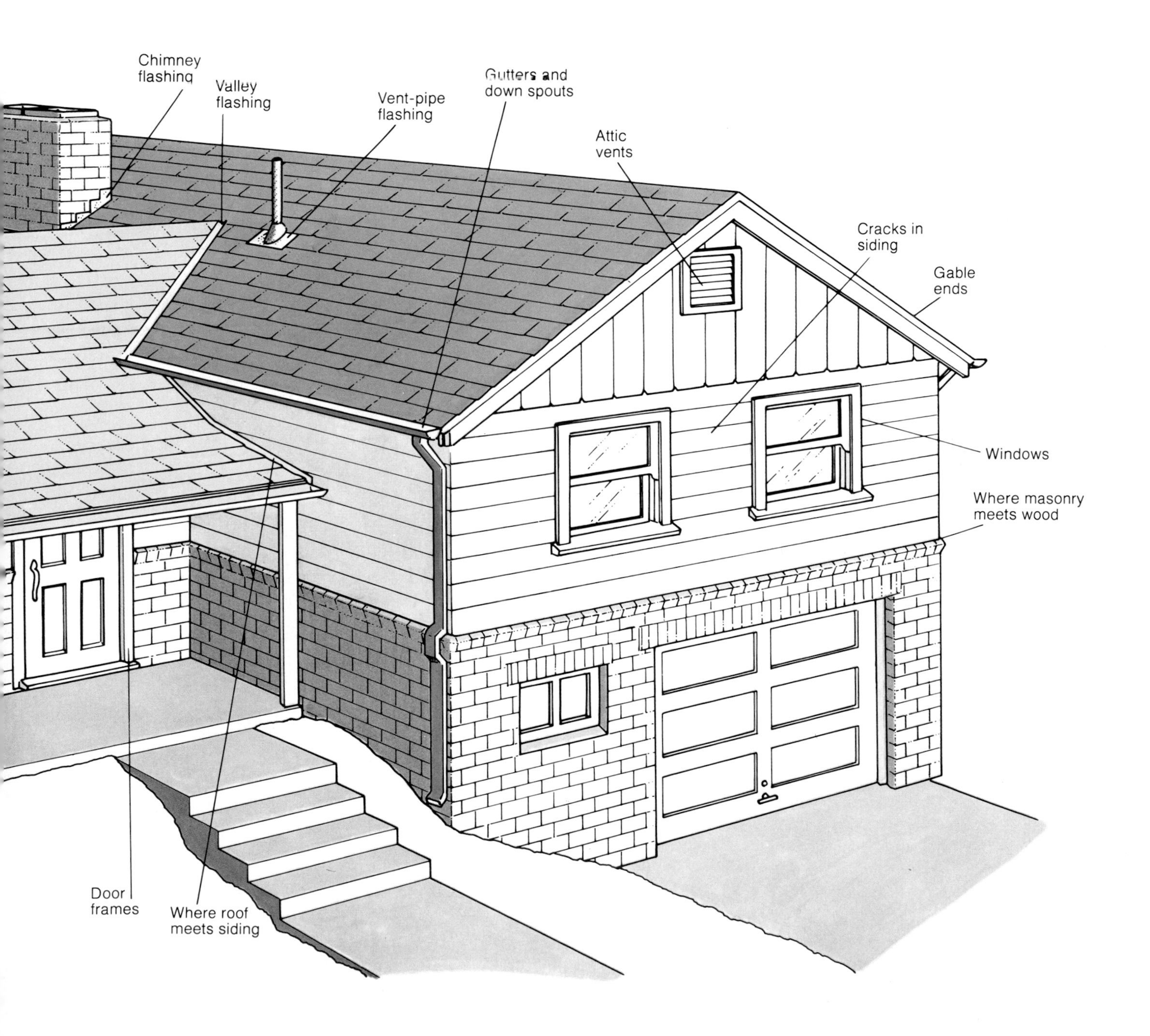

How to Caulk

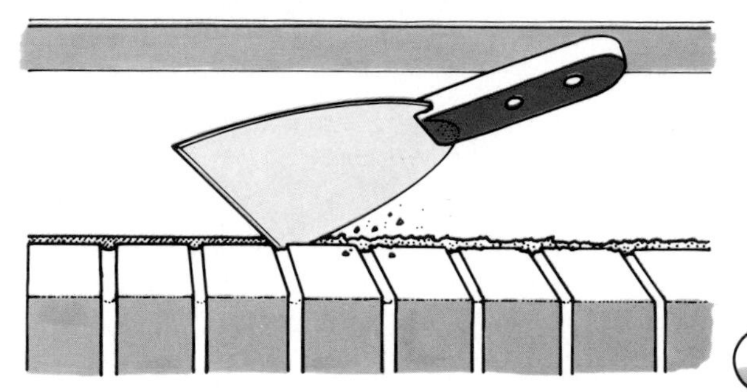

1. Thoroughly clean the crack or joint to be repaired, using a putty knife to scrape away old caulk, paint, or other accumulations of debris. If your caulk requires the crack or joint surface to be primed, apply the primer and allow it to dry. Mask adjacent areas for easier cleanup later. Do not caulk in cold weather (under 45° F).

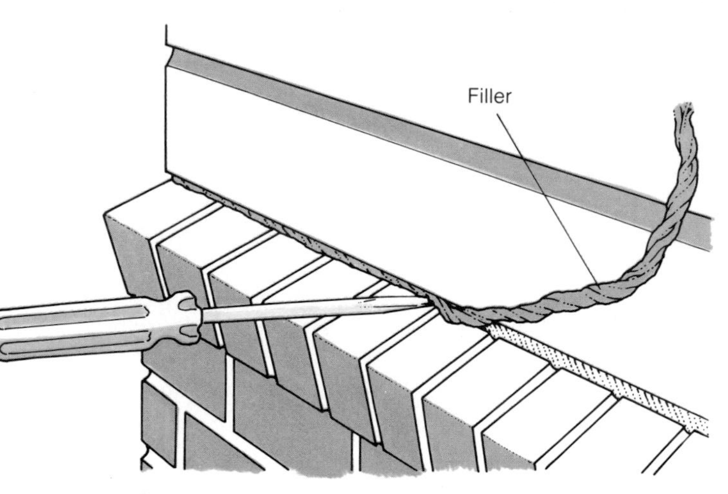

2. If the cracks you intend to fill are larger than about ½ inch, fill them with strips of insulation, sponge rubber, oakum, or other filler before caulking.

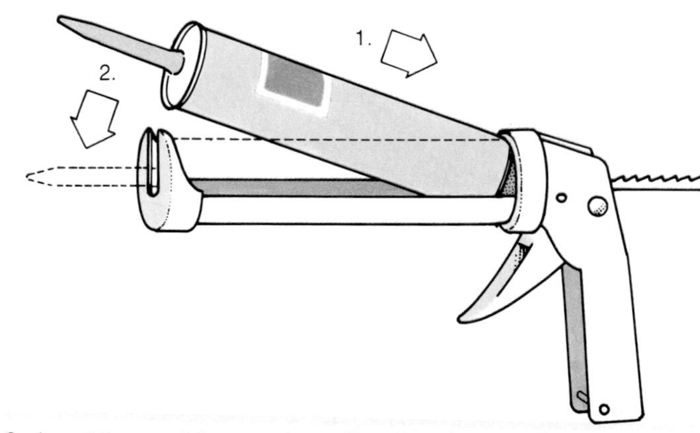

3. Load the caulking gun by pulling the plunger all the way back, turning it so that the notches face up, away from the trigger. Drop the cartridge into the open barrel rear-end first, turn the plunger so that the notches face down toward the trigger, and push the plunger or pull the trigger until the plunger is in position for use.

4. Cut off the tip of the cartridge nozzle on about a 45° angle, and only as wide as the space to be filled (usually no more than ½ inch). The closer to the tip you cut, the finer the bead of caulk will be. Use a long nail or a piece of stiff wire to break the cartridge seal inside the nozzle.

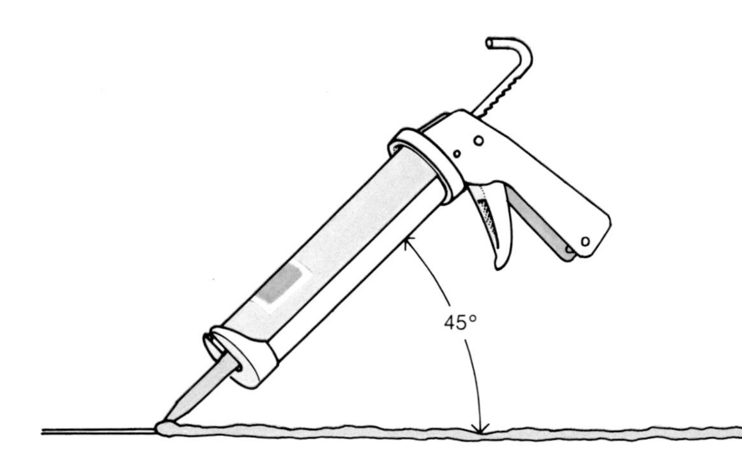

5. Hold the gun at 45° to the surface you are sealing, and maintain a steady pressure on the trigger, forcing the caulk out ahead of the nozzle in order to push it as far into the crack as possible.

Before you begin, practice drawing a steady bead of caulk. Caulking can be difficult at first—even after you have stopped pulling the trigger, pressure on the cartridge continues to release caulk; also, sometimes the caulk will adhere to itself rather than to the surface, and will pull away from the crack or joint along with the bead still spilling from the gun.

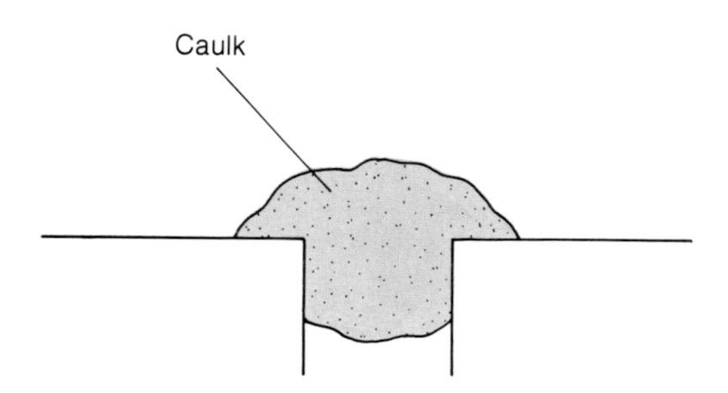

6. Move the gun along slowly and steadily, allowing the caulk to overlap both sides of the crack or joint in order to make a tight seal.

7. When you have finished caulking, loosen the plunger to stop the pressure on the cartridge, and cover the nozzle securely to keep whatever caulk remains in it from hardening. If you insert a broad-headed nail into the nozzle opening before covering it, and wipe off the caulk that seeps out around it, the nail itself will act as a nozzle cap. Cover this with metal foil or plastic.

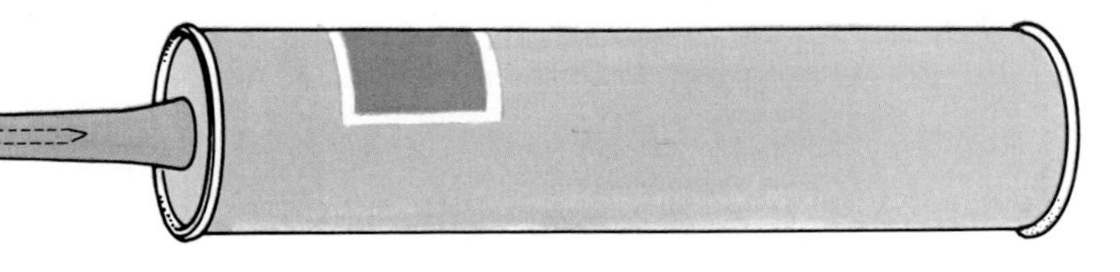

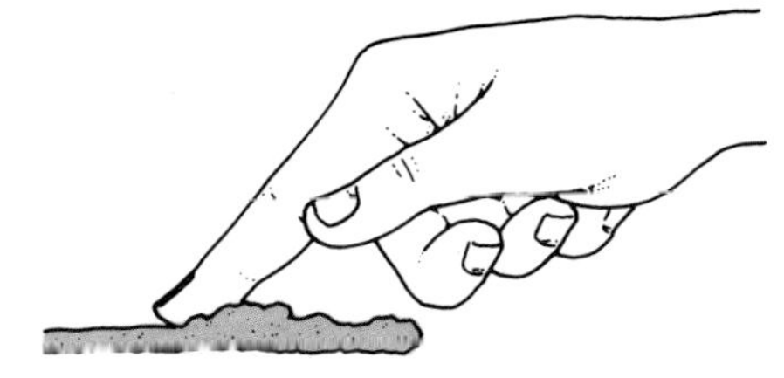

8. Clean excess caulk off your seal and off adjacent surfaces. All caulk can be finger dressed (smoothed over with the fingers) after application—some, such as latex, with soapy water; others, such as butyl rubber, with mineral oil.

Coil Caulk

Coil caulk is simply unrolled and pressed into place with the fingers.

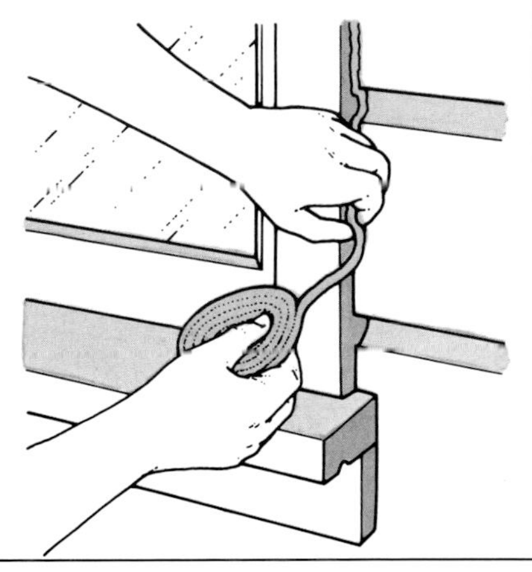

Small Jobs

Small tubes of caulk, used for small repairs such as the cracks behind faucets, also have plastic tips that are trimmed at an angle for proper bead. Apply the caulk as you would apply glue from a similar squeeze tube.

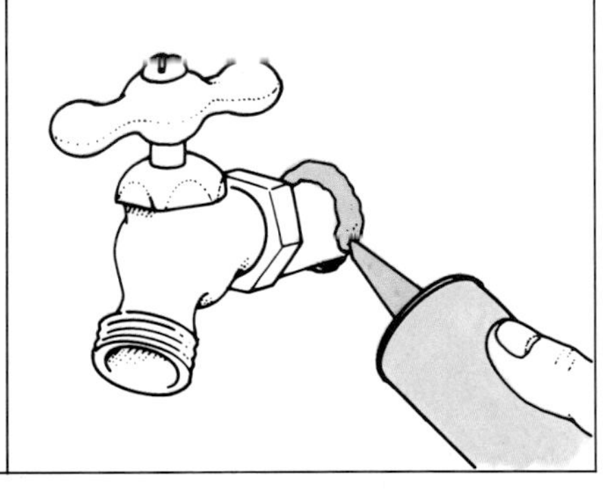

■ **Caulking materials:** No matter what the brand names, specific sealants, forms, or prices, caulking comes in some basic qualities. This chart identifies the ingredients that make up most caulks to help you choose your caulking material wisely. Before buying any caulking compound, and again before using it, *read and follow the instructions on the container.*

If you have specific questions about a particular application and cannot find the answers here, ask the salesperson or someone who has done caulking before.

Caulking Material	Cost (approximate, per 11-oz. cartridge)	Life	Application	Remarks
Oil base	$1-2	1-2 years	All household surfaces.	Least expensive, least durable of all caulks. Oil tends to seep out and may stain unprimed surfaces, while drying out caulk itself, leading to shrinking, cracking, and the need for replacement.
Latex base	$3	3-10 years	Indoors; outside only if painted, and not where seal moves or is subject to moisture. Only for narrow joints (less than ¼").	Because it is water-soluble, should not be used where it will become damp. Insufficiently flexible to withstand frequent expansion and contraction, and should not be used on moveable joints. May disintegrate on concrete or cement.
Butyl rubber	$3	3-10 years	Outdoors, especially on metal and stone; good where water may collect such as eaves and downspouts, since it is water-resistant. Only for narrow or moderately narrow joints (less than ½").	Needs no paint. May take more than a week to cure.
Acrylic latex	$4	10 years	All applications, indoors and out. For joints up to ½" × ½".	Durable, fast-curing (about 1 hour), does not stain. Should not be painted.
Elastomer (synthetic base: silicone, polysulfide, or polyurethane)	$6	20 years +	All applications, but does not adhere to paint.	Highly flexible, long-lasting caulk. May be used on large joints. Some require surface priming before application. Cannot be painted.
Lead base	—	—	—	Known to be toxic; illegal in many states. Use is discouraged.

Insulating Windows

Windows, including the glass panes in doors, lose energy in three ways. First, they allow air to pass around their frames and edges; second, they allow heat to pass through the glass itself, in the process of conduction; third, heat can escape just by "shining" out—radiating to the cold outside. You can stop the first kind of loss by caulking and weatherstripping, the second kind by double or triple panes and storm windows, and the third kind by drapes, shades, or other window coverings.

If you stand beside your bedroom window on a frosty winter morning, you'll notice that cold air from outside is leaking into your house at what could be considered an alarming rate. You can actually feel the cold seep in, chilling the area around the window as it falls toward the floor (cold air is heavier than warm air). But a few feet in from the window, the air is warmer and more uniform. What happened to all that cold? You paid to heat it.

Reports published by the Texas Energy Extension Service and other investigative organizations indicate that a single pane of clear glass may lose as much as *20 times* more heat than a well-insulated wall.

If you look around your house at the enormous square footage occupied by very loosely covered holes in the wall (i.e., windows), you can begin to appreciate why your utility bills are so high. And not only do you lose warm air in the winter; you also lose cooled air in summer. So those holes in the wall consume your energy dollars year-round. In fact, *any time* the temperature inside your house is different from the temperature outside, heat is passing through your windows *in whichever direction costs you money*!

This doesn't mean that you should board up your windows—you want the sunlight they let in, you want to be able to see the outside world, and frequently you even want some air. But it *does* suggest that you should take a look at how you can have all the benefits of windows without spending so much money on their energy-eating habits.

For simplicity's sake, let's just deal with the problem of losing heat in cool weather. When you conserve heating energy at your windows, you will be setting the stage to conserve cooling energy at the same time.

Incidentally, in examining some of the causes for your high energy bills and exploring ways to lower them by reducing your home energy consumption, don't neglect the little items. It's amazing how many of us who are earnestly concerned about our use of energy don't recognize the importance of such things as the window that was cracked five years ago that no one has ever gotten around to replacing; the window that never shuts tightly because of a loose latch; the long winter minutes a loving mother spends shivering in an open doorway, watching her child trudge off to school through the snowdrifts, while the furnace labors to replace the lost warmth. No amount of caulking, weatherstripping, or other home amendments will reduce your energy consumption if you periodically toss what you have saved out the window.

(To replace that broken window, see page 29 of Ortho's book *All About Home Repairs*.)

On February 9, 1977, a front page story in the *Wall Street Journal* indicated that when the outside temperature was 0° F, the temperature on the inside surface of a single-pane window, in a room heated to a normal household temperature, was a chilling 17°. By merely adding glass or plastic storm windows, the temperature on the inside surface of that window was raised to 47° F; and by the further addition of a single sheet of plain freezer wrapping, the inside surface temperature of the window went up to 60° F.

Does insulation work? You bet your energy-saving dollar it does. What you have to decide is whether it is cost-effective in your own case—whether the expense of keeping your window warm can be justified by the energy savings in a short enough time to warrant the outlay.

■ **Storm windows:** Windows can be insulated in a variety of ways, but the most common and traditional window insulating amendment is the exterior *storm window*.

The term "storm window" often is used loosely to mean *any* sort of window insulation. Actually, however, a true storm window is a wholly independent sash. It may be entirely removable or permanently installed, but in either case it is separate from the primary window. For our purposes, we will define a storm window as a single layer of clear material that is used to insulate a window, whichever method is used.

■ **Storm doors:** Like the true storm window, the traditional storm door is not a material added to an existing door, but rather is a whole, separate unit designed to provide the same sort of extra outside protection for the doorway that storm windows provide in their place. Ordinarily, a storm door is made of a solid material such as aluminum on its bottom half; the top half or two-thirds of the door is glass, held in an aluminum frame. Some storm doors have tracks that allow the glass to slide away leaving a screen for summer cooling. The doors are easy to put up or take down, by removing the standard pins.

Plastic Sheeting Storm Windows

One of the simplest forms of exterior insulation is the plastic sheeting storm window. It is made from 6-mil polyethylene plastic. You can buy it at most hardware stores or from mail-order catalogues. When sealed tightly around all its edges, this extremely inexpensive form of makeshift storm window (about 70¢ per linear foot in 10-foot widths, or about $35 to $45 for a roll 10 by 100 feet) is essentially as effective as a far more complicated and expensive one made from glass or rigid plastic, approximately doubling the insulation efficiency of a single-pane window. The reason is that no glazing material in itself provides much in the way of insulation—what it does provide is the extra wall that holds the additional layer of air, the *real* insulation.

Polyethylene sheeting has its disadvantages, however: it prevents you from opening your window while it is up; it cannot be reused because it tears, wears, and yellows, even when transparent (and treated not to go opaque) instead of the more commonly sold opaque white; it distorts your views; and it is not particularly attractive.

The following instructions are for installing the sheeting on the outside of your windows.

1. Outside your house, measure your largest window from the outside of the frame on one side to the outside of the frame on the other side. When you buy your roll of plastic, make sure it is at least that wide so you can cover your entire window.

2. Measure the height of *all* your windows, from the outside of the top of the frame to the outside of the bottom of the frame. Add the total heights together

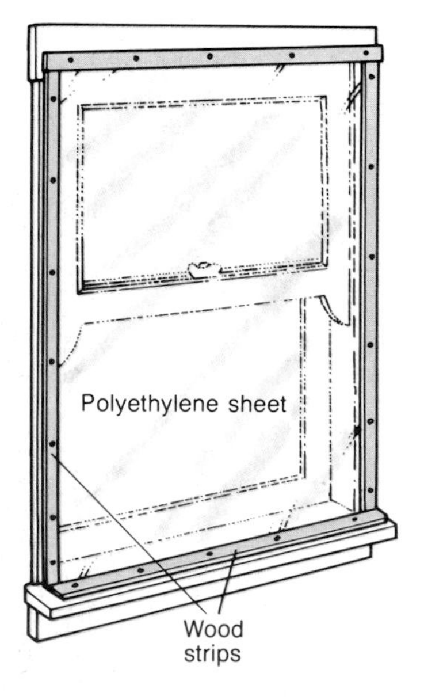

so you'll know how long a roll, or how many rolls, you will need. Add 2 inches of height for each window as a safety precaution (unless you really like going out to the store).

3. When you buy the plastic sheeting, also buy ¼ by ½ inch wooden strips, or lengths of thin molding, to match the length and width of every measurement you have taken.

4. Cut the plastic sheets to cover each window around the outside of its frame.

5. Starting with the top edge of each window, roll the vinyl around the wooden strip to ensure a tight fit, then nail through the wooden strip to hold the plastic to the window frame so that the nails don't tear it.

6. Keeping the plastic as taut as possible, nail or staple around the entire window frame.

7. Trim off any excess plastic or strips of wood. You can paint the wood, if you wish.

■ How heat moves through glass: Not all windows are alike. Generally speaking, a window is a sheet of glass set into some sort of frame that fills an intentionally created hole in the wall. That glass commonly comes in three weights: single pane, double pane or double glazed, and plate glass. The different weights possess different strengths and need to be evaluated in terms of their tensile qualities as well as their ability to conserve heat. Use single-pane glass for a window up to about 5 square feet. After that, single-pane glass is no longer really safe—use double-pane glass instead. If the surface of the window is about 10 square feet or more, use plate glass, which is heavier and thicker.

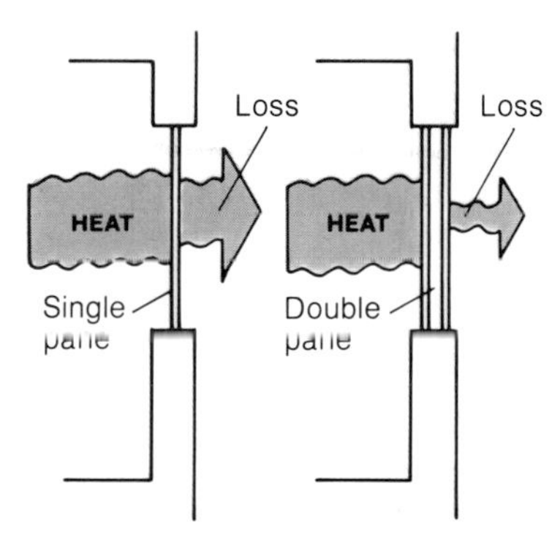

Glass is an excellent heat conductor; but when that glass is functioning as a window, its ability to conduct heat is not a desirable feature. Motionless or nearly motionless air, on the other hand, conducts heat very poorly. Therefore air is a good insulator when it is in a space small enough to keep its molecules motionless.

The principle behind both thermal windows and storm windows is to create virtually motionless *dead air spaces* between layers of glass. (Thermal—double- or triple-glazed—windows have all their layers of glass in a single sash. Storm windows, discussed on page 34 use more than one sash to hold multiple layers of glass.) Since moving air is a good conductor of heat, the air must be fairly still; therefore, the space between the layers of glass must be sealed tightly. That space should hold as much of the insulating air as possible, yet it must not be too wide or else the air will begin to circulate. Most double-glazed windows have an air space that's between 3⁄16 and ¾ inch thick; storm windows may be more than an inch away from the interior glass. Anything up to about 3 inches will improve a window's insulating capacities enormously.

One effective way to improve a window's insulating ability is to increase the number of layers of dead air. This works better than increasing the thickness of a single layer, which would allow the air to move and to conduct heat before it got thick enough to really increase its insulating capacities. To increase the numbers of layers of air, a second—and sometimes even a third—layer of glass is added.

Glass is heavy and hard to work with, however, and our modern age offers a few technologically advanced alternatives, one of which is plastic. While mylar and tedlar films have life expectancies of several years, rigid plastics generally are superior to plastic films, because they are sturdier; they withstand more abuse; they last longer; and in the long run, they are less expensive. It's true that most films will pay for themselves in a single year by saving energy; however, they will last only that year, and you will need to buy new film the following winter. While good-quality rigid plastic may take 5 to 8 years to return its cost in energy savings, it can last as long as 10 to 15 years. However, be sure to use *high-quality* rigid plastics such as acrylics; many plastics of inferior quality will yellow or become scratched over time.

In commercially produced thermal windows, the space between the layers of glass or plastic is often filled with a moisture-free gas or has desiccant granules added. The desiccant absorbs the moisture that's left between layers when glass is sealed at the factory, and prevents condensation from forming in the middle of the window.

If you plan to add a layer of plastic to your windows, or if you purchase unsealed double-glazed windows, make sure that they have tiny vents (sometimes called "weep holes") in the outside pane to allow moisture to escape. In addition, be sure to seal the inside pane tightly.

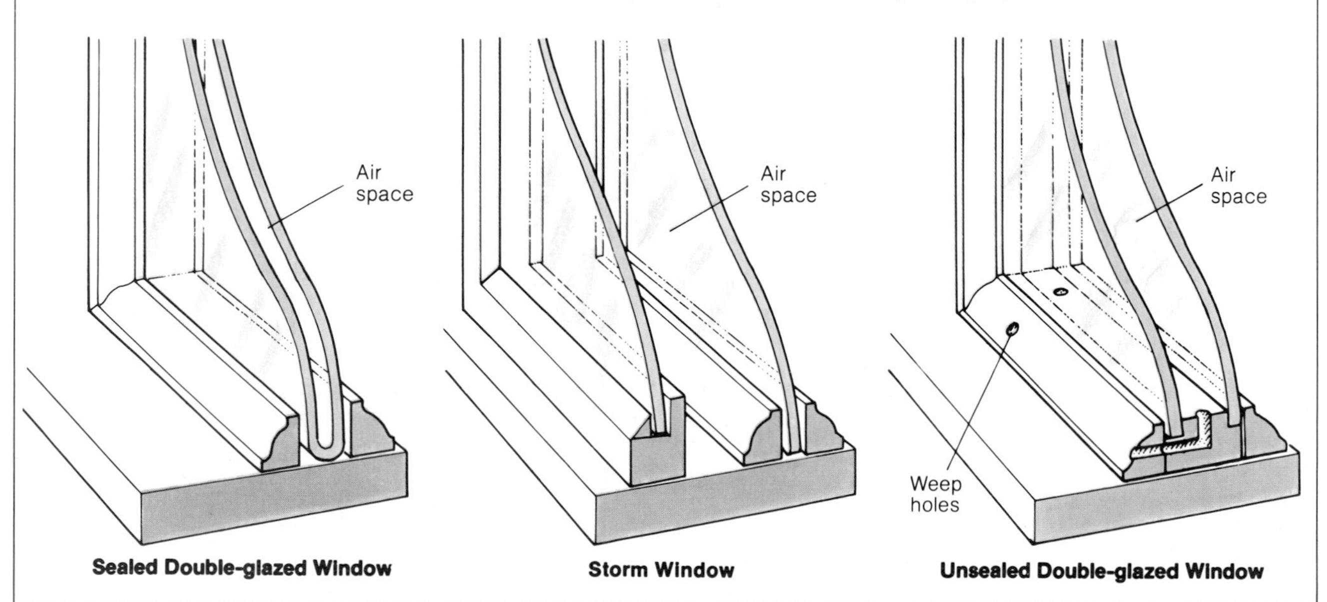

Sealed Double-glazed Window **Storm Window** **Unsealed Double-glazed Window**

Glass Storm Windows

While plastic sheeting storm windows have distinct advantages, most are too fragile to last more than a year. For greater durability, consider rigid acrylic storm windows (see below), which may last as long as 15 years indoors (fewer outdoors, where conditions are more extreme). But for insulation that will last the life of your house, and will approximately double the insulation efficiency of a single-pane window, install that longtime cold-climate favorite, glass storm windows.

Removable glass storm windows are the familiar single-pane outdoor windows that you put up each fall and replace with screens each spring. Often you can buy them ready-made to fit your window size. A standard 3 by 5 foot single-pane removable storm window in an aluminum frame costs about $30 to $50. You can also have these windows made to order; the expense probably won't be much greater.

Permanent glass storm windows also are made to be installed outdoors. They come as single panes and occasionally as double panes, and have moveable screens built into the same sash for year-round use. In warm weather, you just slide the storm glass out of the way, and slide the screen in place. Like removable storm windows, permanent ones may be purchased ready-made (about $65 to $80 for a 3 by 5 foot window), or be made to order. Permanent storm windows are somewhat more expensive than removable ones, but they are more convenient. If you install them yourself, you may save a fraction of your purchase price—about 10 percent—but the effort is considerable for the money you save. It's much easier to have them installed by a contractor.

For either kind of window, consult your Yellow Pages under *Windows* or *Storm Windows*.

Other Types of Exterior Window Insulation

There are other ways to insulate your windows besides adding layers of air between sheets of transparent glass or plastic.

For the sake of clarity, we have divided window amendments into two general categories—interior and exterior. In general, exterior amendments make less effective insulators than interior amendments because it is difficult to create a tight seal with most of them. But in combination with double glazing and/or storm windows, they can be quite effective. And there is no reason why they can't be used with, rather than instead of, interior amendments. With some amendments, like window treatments, home energy conservation projects may build upon one another: If a single project will save you $3 per month and another will save you $6 per month, doing the two together may well save you more than $9 per month.

At the same time, consider the true usefulness of mechanical devices. For instance, the installation of rolling shutters may cost you more than the shutters will save you in heating.

■ **Wooden shutters:** Depending on the kind of wood, a solid shutter will approximately double the insulating value of your unprotected single-pane window, at a cost of about $2 to $3 per square foot. As with foam boards, tight fit is sometimes a problem.

■ **Wooden louvers:** Wooden louvers cost about twice what solid wooden shutters cost, and are not quite as effective as insulators. Their advantage, however, is that they are easily controlled to admit air and sunlight.

■ **Insulated aluminum louvers:** Manufactured for industrial purposes but equally useful in your home, insulated aluminum louvers cost about $10 per square foot, and provide insulation comparable with that of the rolling shutters described below.

■ **Rolling shutters:** Sold under a variety of brand names such as Rolladen and Roll-Awn, these attractive shutters—long popular in Europe—cost between $2 and $8 per square foot in materials, and will approximately triple the insulating value of your unprotected single-pane window. They are made of wood, hollow plastic, or foam-filled plastic interlocking slats that roll up to store in a box above the window. When you want to use them, just pull them down with a rope, crank them down, or release them electrically with a button. When completely closed, they form a double air space—one between the window and the shutter, and another between the outer and inner sides of the slats. Because they are guided in tracks or channels on their ends, they can achieve a tight closure over your window. Left loosely closed they provide effective shading on south-facing windows, permitting some light but little heat to enter a room through the slats.

Wooden Shutters

Rolling Shutter

Insulated Aluminum Louvers

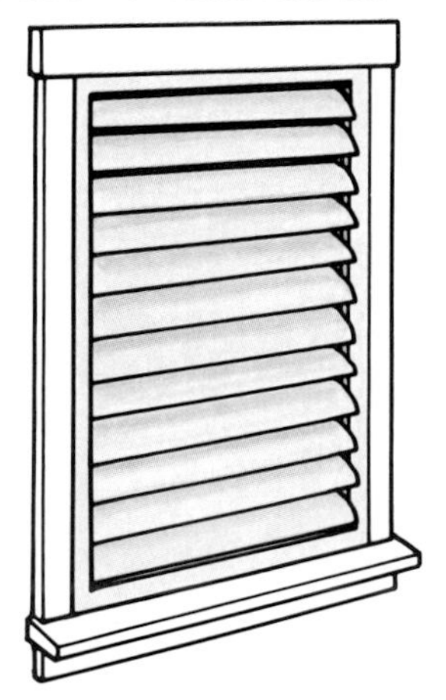

■ **Shading:** Some exterior amendments conserve energy by keeping the sun

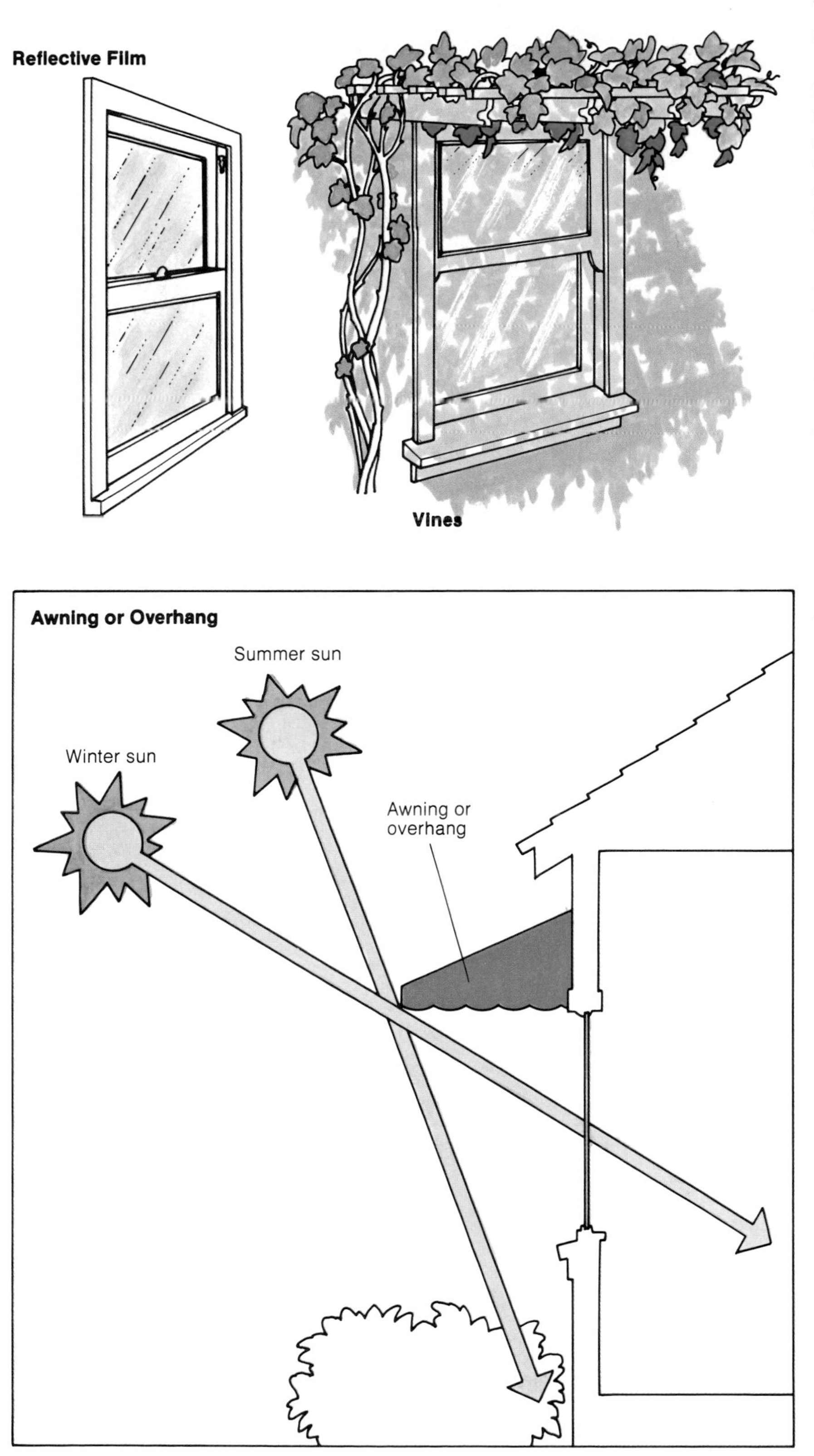

out of your house in hot weather, thereby reducing the cooling load.

Vines, trellises, trees. Deciduous trees at south-facing windows are particularly useful—they admit sunlight in the winter when they have lost their leaves, but block it out with their foliage in the summer, keeping the house cooler. See pages 46-48 for further details about landscaping for energy conservation.

Awnings, overhangs. Like vines, these are particularly helpful at south-facing windows. They can be built to maximize winter sunlight and minimize summer sunlight according to the latitude at which you live (see "Solar Energy," page 69, for ways to use the sun).

Interior Window Insulation

If you're building a new house, why not take the opportunity to install double-glazed windows—and storm windows as well, if you live in a cold climate. And there's nothing to stop you from adding some exterior insulation while you're at it—you'll recoup your investment in both comfort and energy savings for appropriate use of all these amendments.

Whether you're building a new house or improving an existing one, the most cost-effective single unit of window insulation you can install will be *inside*. And if you're remodeling, interior window insulation will also be more cost-effective than any other window amendment except outside shading.

Interior window insulation comes in a broad range of materials, from an interior plastic storm window, or even a simple roll-down window shade, to multilayered shades and curtains.

Any material or air buffer you can place between the window and the air in your house will cut your energy losses somewhat. The basic principles of interior window insulation are the same as those of exterior window insulation: Keep conditioned air in and unconditioned air out. Therefore, some of the forms of interior window insulation are similar or identical to some of the forms of exterior window insulation. But because the inside is not directly subject to wind, rain, snow, and other exigencies of nature, more fragile materials may be used. Often, these are not only superior to the tough exterior materials needed to brave the weather, but also nicer to look at.

One of the attractions of storm windows is that they provide insulation without blocking sunlight or your view. But at night when you are sleeping, you don't much care about sunlight or views. Therefore, to insulate your windows from within, you might consider one of the many forms of window insulation that you can open or close with varying degrees of ease. For example, roll-up screens or window shades simply pull down or snap up with the flip of a finger. Louvered blinds can be opened at an angle to act as passive solar energy collectors (see "Solar Energy," page 69).

Drapes that extend the entire height of a window can be effective window insulators while also fulfilling a role as interior decoration. The best drapes for this purpose will be lined, and made of a heavy material. They should fall completely to the floor at the bottom, extend all the way around the sides of the window, meet at the top, preferably with a valence. The more airtight the drapes are, the more effective they will be.

Acrylic Plastic Storm Windows

You can make a single-pane, solid plastic interior storm window yourself. And although it is somewhat more complex and expensive to construct than plastic sheeting storm (see page 32), it also is reusable, relatively attractive, and less fragile than either glass or 6-mil polyethylene. The expense of this project is limited to a few screws and some ¼ inch clear acrylic plastic, at $2 to $2.50 per square foot.

1. Measure the length and width of the window you intend to cover, so that all the edges of your storm window will extend 1 inch over a flat surface. The window casing will work well if it is not a curved molding.

2. Draw the storm window you just measured onto masking paper covering a sheet of ¼ inch clear acrylic plastic.

If you are adept with a circular saw, cut the acrylic yourself. Otherwise, ask your plastics supplier to cut the acrylic to your *exact* measurements, or to have it cut for you. Do not practice on this sheet.

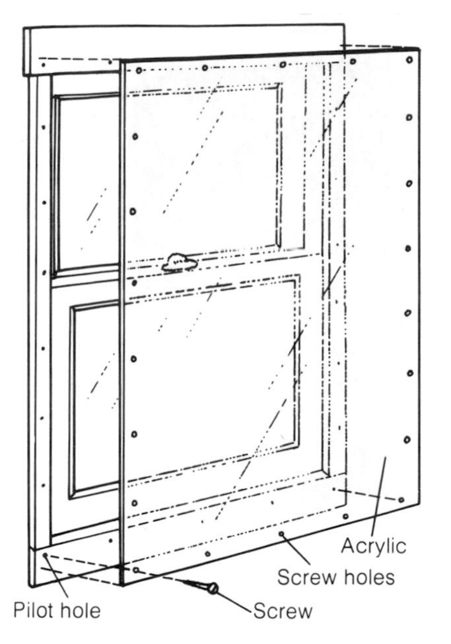

3. Sand the edges of the acrylic with fine sandpaper. Starting with the four corners, drill holes evenly spaced, no more than 12 inches apart, ½ inch in from the edge of the acrylic.

4. Hold the acrylic over the window it is to cover, and mark the screwholes on the window casing or other flat surface surrounding the window.

5. Using a smaller bit than was used to drill the acrylic, drill starter holes in the casing where you have made marks.

6. Hold the acrylic up to the holes and screw it to the casing with brass-finish or roundhead screws.

7. If the acrylic doesn't fit tightly, insert felt, reinforced felt, or plastic foam weatherstripping (see page 22).

You can remove this storm window any time simply by taking out the screws.

If you store the storm window, cover it with soft cloth so the acrylic doesn't get scratched.

Other Interior Window Insulation (see chart at right)

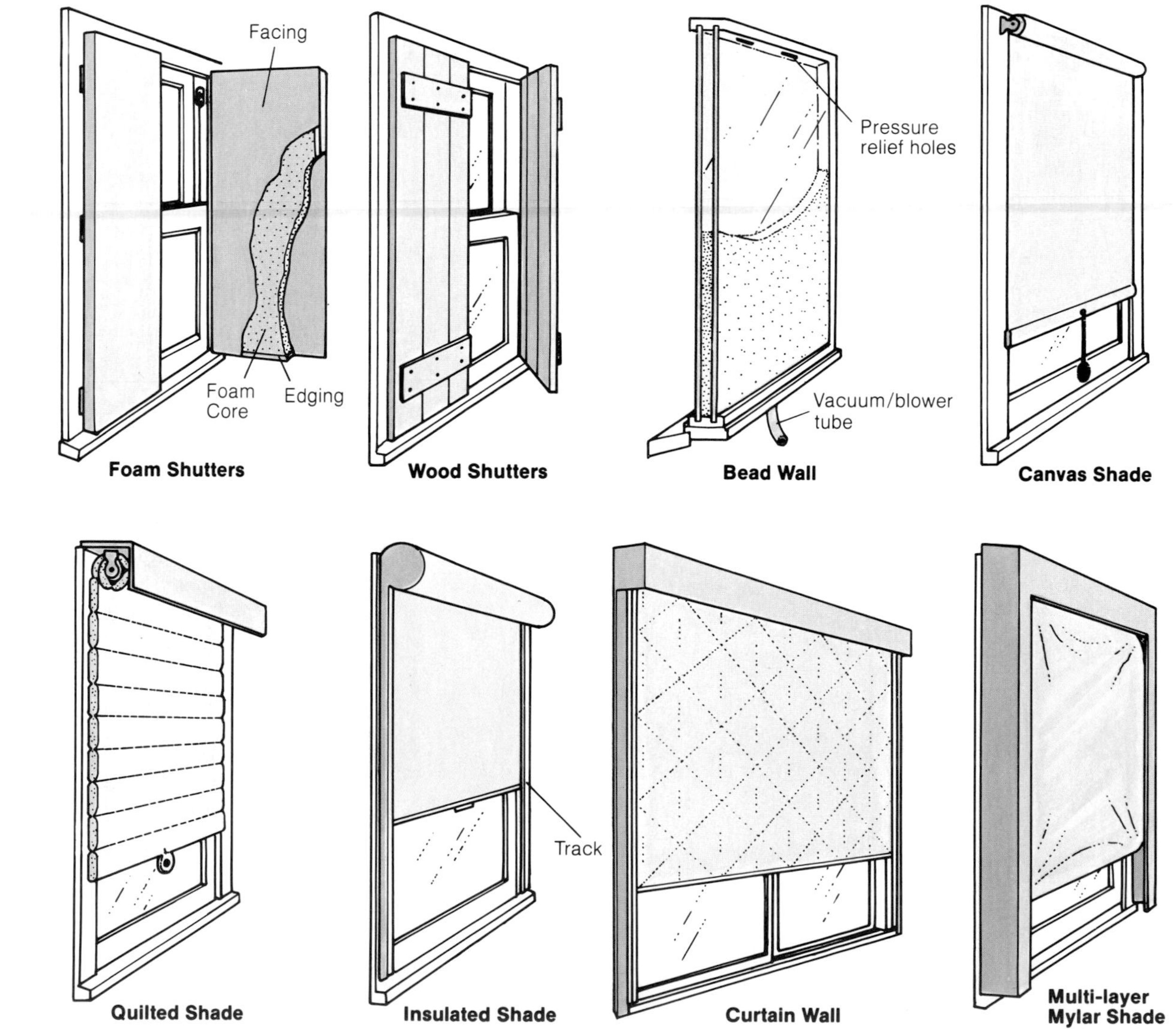

Types of Interior Window Insulation

Insulation	Cost (materials only)	Effectiveness	Remarks
Foam boards (1" thick) (insulating shutter)	About $1 per square foot	Very good. Improve single-pane insulation about five times.	Sometimes difficult to achieve tight fit; not particularly attractive without additional covering. May want to frame with wood, then hinge, and fit to window. Most foam boards are made of flammable materials, so face with aluminum or some other fireproof material.
Wooden shutters	$2-3 per square foot	Acceptable. Double insulating efficiency of single-pane window.	Solid shutters are more effective than louvered ones. Sometimes difficult to achieve tight fit.
Reflective film	50¢-$1 per square foot	Acceptable. May double insulating efficiency of single-pane window.	Easy to apply in small areas (up to about 2′ × 3′), but reduces incoming light and may blister, peel, and look unsightly.
Canvas roll-down window shades	$2-6 for 3′ × 6′ shade	Better than nothing. Cut heat loss through the window by about 25%.	Easy to put up, easy to use, and if you don't like their looks you can paint or cover them. Of minimal insulation value—they don't make a tight seal anywhere, and wind and cold can penetrate them.
Acrylic plastic storm window	$2-2.50 per square foot for ¼" thickness	Acceptable. Doubles insulating efficiency of single-pane window.	Available in kits at some home improvement centers, or can be made at home (see instructions, p. 36).
Quilted shade	About $4 per square foot for five layers of insulation	Good. Improves insulating efficiency of single-pane window about four times.	An insulation sandwich with reflective Mylar at the center, surrounded by two ¼" thick layers of polyester batting, surrounded in turn by two layers of polyester-rayon. Tracks at the sides and weights at the bottom ensure a tight seal.
Insulating shade	About $6-7.50 per square foot	Good. Improves insulating efficiency of single-pane window about four times.	Three shades—clear inside, tinted in the middle, aluminized reflector on the outside—guided on tracks to ensure tight seal, and designed to make use of solar heat in a passive fashion as well as to insulate.
Curtain wall	About $4-7 per square foot	Excellent. Improves insulating efficiency of single-pane window about ten times.	Four layers of metal fabric curtain designed to be guided by tracks (not included) for tight side seals. When not in use, the curtain wall stores on a roller and is lowered and raised by an electric motor or by hand. Intended to cover large window areas only.
Multilayer Mylar shade	About $4 per square foot. Frame about $1.50 per linear foot; storage unit about $3 per linear foot	Remarkable. Improves insulating efficiency of single-pane window about 15 times.	Five layers of aluminized mylar expand to 3½" of baffled air spaces when drawn. Frame with tracking must be bought or built separately to create tight seal. Maximum width 64", maximum length 8′.

Insulating the House

You've caulked the cracks and weather-stripped the windows. The coldest air can't crawl across your threshold. At every window, panes of glass combine their strength to beat back infiltration.

If yours is a typical American house, your heating and cooling bill has already plummeted—10, 15, maybe even 20 percent. And yet, as you sit before your energy-efficient wood stove (see "Wood and Other Alternatives," page 95), you feel . . . cold air!

It's coming through the walls!

Your house itself needs insulation. How much depends on where you live, the structure and site of your building, and other factors. But even if you already have insulation, the odds are great that it isn't enough to keep out the winter cold and summer heat.

Until very recently, it cost so little to heat and cool our homes that it hardly made sense to maintain efficient insulation, and houses were therefore constructed with little of it. However, for the present and the foreseeable future, insulation is one of the best investments you can make. You might say it's a hot item.

It would be self-defeating to insulate without first caulking, stripping, and protecting your windows and doors. If you neglect these crucial steps, insulation alone won't keep whatever heat or cooling might be saved from slipping away into the great outdoors. However, once you have sealed the various seams and edges of your house, you can save an *additional* 20 to 25 percent on your space-conditioning bill by bringing its shell up to the same effective standards.

■ **Insulating materials:** What constitutes a good insulator? Wood, brick, plaster, masonry, and glass—all materials that are commonly used to build a house—are not particularly effective insulators, although they *store* heat well. Rock wool, fiberglass, polystyrene, and urethane do the job much better.

An effective housing insulation should have these characteristics: it should be moisture, fire, and rot resistant; it should not attract feeding or nesting mice or other vermin; it should not emit unpleasant odors; and it should not be so heavy that it strains the structure of the building. But the most important quality of all is: it should have the ability to *resist the conduction of heat.*

The ability of any material to resist heat conduction is known as its **R-value**. The greater a material's R-value, the more efficient its heat-resisting qualities; therefore, the better it can insulate.

Because *air itself* is such a good insulator, the best housing insulation materials generally trap many minute pockets of air in a compact space. As a result, housing insulation tends to be bulky without being very heavy.

A material's R-value (and therefore its quality as an insulator) is calculated on the basis of its thickness as well as its composition. R-value is the criterion by which housing insulation is sold and applied, rather than according to its weight, density, thickness, or other qualities. For example, R-19 is the insulation recommended by the Federal Housing Administration, the American Society of Heating, Refrigeration and Air Conditioning Engineers (ASHRAE), and other insulation research and manufacturing organizations for the outside walls of most houses in the United States.

The chart below indicates what R-19 insulation means in terms of several common insulators and several common building materials.

Thickness of Various Insulating and Building Materials Required to Reach R-19

Material	Required Thickness (approximate)
½" Plywood	18" (1½')
Plasterboard	24" (2')
Concrete cinder blocks	72" (6')
Common brick	96" (8')
Sand	156" (13')
Fiberglass batts	6"
Cellulose fiber batts	5"
Urethane foam	3"

■ **Vapor barriers:** The air in your heated, insulated house contains a surprising amount of water in the form of vapor. As the warm air attempts to pass through your ceilings, walls, and floors to the cold outside, it carries this water with it. When the water starts to penetrate your building's shell and reaches cooler air, it condenses. Then it collects in your insulation and other nearby materials, such as wallboard and structural timber. This makes the insulation ineffective; and, along with the other affected materials, it begins to deteriorate.

To protect your building, as well as your insulation, from moisture, a **vapor barrier** is recommended with virtually all insulation. This is a layer of water-repellent material—for example, aluminum foil or thick plastic sheeting—taped at all seams with a water-repellent tape. It lies on the warm side of the insulation (*toward* the inside of the house, not away from it), between the warmth of an inner wall and the cold temperature just beyond it.

Most insulation sold in the form of blankets or batts comes with an attached vapor barrier. And some of the extruded insulations act as their own vapor barriers. But if you are installing loose fill, rigid boards, or some sort of blanket or batt that has no vapor barrier, you need to create your own. Aluminum foil or plastic sheeting and waterproof tape work well. So does painting at least two coats of an oil-based enamel paint on the inside walls in front of the insulation (or on the floor above it, or on the ceiling below it).

■ **Installing a vapor barrier:** If you plan to install your own vapor barrier, consider polyethylene sheeting—it's easier to buy and use than large rolls of aluminum foil. The plastic should be at least 2 mils thick.

For an unfinished wall, drape a sheet of polyethylene around the entire wall, on the inside of both insulation and studs. For an unfinished floor or ceiling, follow the same procedure or, cut strips of plastic to fit between joists, allowing about 4 inches extra width to lap up the sides of the joists. Staple these in place as smoothly as possible, then tape any gashes in the plastic with waterproof tape.

Remember to keep the vapor barrier toward the warm side of the insulation—the *in* side of the house.

Making Your Own Vapor Barrier

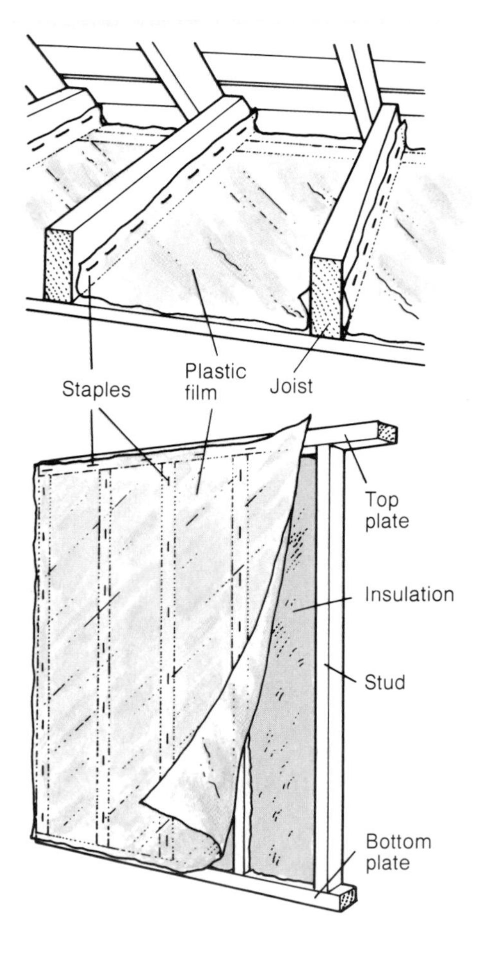

Forms of Insulation

Insulation materials are packaged and sold under many brand names, sometimes with minor variations in composition. However, only nine materials are currently used for housing insulation, and these materials come in only five basic forms. This chart tells you how these five forms are generally used, what materials they are made of, and how easy or difficult they are to handle.

You will also want to know the R-value of the insulation. This depends on the material it is composed of, and details about each of the materials listed here are included in the chart on page 40. The R-value is usually printed on the face of the insulation or, in the case of loose fill, on the bag.

Form	Materials	Principal Use	Remarks
Loose fill: Pebbles, pellets, or shredded paper. Easily poured into flat, open spaces such as unfinished and unfloored attics, or blown with air jets into covered walls or other irregular cavities.	Perlite/ Vermiculite Fiberglass Rock wool Polystyrene Cellulose fiber	Floors, walls	Very easy to apply, and fairly commonly used by people doing their own insulating. Has a tendency to settle, particularly if installation is high or slanted, which diminishes its effectiveness by leaving top portions of fill empty or packed far too loosely. Vapor barrier should be installed separately.
Blankets or rolls: Sheets of insulation several inches thick, 16′ to 64′ long, and 16″ or 24″ wide (to fit easily into standard stud spacing). Ideally suited for long, unobstructed, accessible spaces not blocked by plumbing, wiring, or other fixtures.	Fiberglass Rock wool Cellulose fiber	Floors, attics, roofs	The form most commonly used by people insulating their own homes. Usually sold with flanges designed for easy stapling, and sold with or without vapor barrier attached.
Batts: Blankets precut to 4′ or 8′ lengths.	Fiberglass Rock wool Cellulose fiber	Floors, attics, roofs	Easier to handle than blankets, but otherwise identical.
Rigid boards: Precut slabs between ½″ and 1″ thick, and from 8″ × 8″ to 4′ × 12′ in area. Best for upright installation such as sheathing under aluminum siding, and basement or other foundation walls.	Fiberglass Polystyrene Urethane Isocyanurate	Unfinished walls	Plastic boards are common, and are highly flammable; they should be covered with fire-retardant wallboard. Some are not water resistant, and some are sold with a bonded aluminum vapor barrier; but the absence of a vapor barrier does *not* imply that the board is water-resistant. Ask your salesperson.
Foam-in-place: Plastic foam shot through a pressurized hose into finished walls between studs, and other hard-to-reach places.	Urea formaldehyde Urethane	Finished walls	Very effective, but very difficult to install satisfactorily. Should be installed only by a reputable, qualified contractor. May shrink, emit toxic fumes, or even swell and burst through walls.

Insulation Materials

As we've said, there are nine materials currently in use for housing insulation. These materials are described in the chart below. But you should not be discouraged by this diversity. You will not have to choose among them all. First of all, unless you are a professional in the building trades, you probably won't even be able to *find* several of the materials listed. Second, if you do find them, you might well decide they are too expensive, or too complicated to install, for your purposes. We have provided more complete information than you are likely to need simply to let you know the lay of the land. If you know that isocyanurate exists, you *may* decide that it suits your needs perfectly.

The materials you are most likely to encounter when you go to your friendly neighborhood home improvement center are, in general, the least expensive, easiest, and safest materials to handle. You will almost always find fiberglass in batts and blankets; you will usually find cellulose fiber as loose fill; and you will often find rock wool, particularly as loose fill. If you are seriously looking for it, you can also track down polystyrene as rigid boards. You can locate perlite and vermiculite easily in plant and garden centers, since both materials are used to lighten plant soil. And if you arrange to have a contractor blow foam into your walls, he will normally use urea formaldehyde.

The R-value for each of the materials varies from form to form. It also varies in another way: A three-inch fiberglass board does not yield three times the insulation provided by a one-inch fiberglass board. However, these numbers, representing the R-values per inch, provide you with the relative qualities of the materials. As we've noted, the actual R-value will be printed on the insulation, so you will know exactly what you are getting.

Material	Forms	R-value Per Inch	Remarks
Perlite	Loose fill	2.3	Easy to use, but least efficient of all standard materials; moisture-retentive.
Vermiculite	Loose fill	2.3	
Fiberglass	Loose fill	2.7	Fiberglass blankets and batts are the forms of insulation most commonly installed without contractor assistance. Easy to handle; both fiber- and moisture-resistant. Wear gloves, long-sleeved shirt, respirator and eye protection when handling, to keep material from irritating skin, lungs and eyes.
	Blankets	3.0	
	Batts	3.0	
	Rigid boards	4.0	
Rock wool	Loose fill	2.7	Similar to fiberglass.
	Blankets	3.6	
	Batts	3.6	
Polystyrene	Loose fill	3.0	Polystyrene boards are commonly used to insulate unfinished walls. Should be faced with gypsum or similar wallboard since polystyrene is highly flammable. Moisture-resistant.
	Rigid boards	3.5	
Cellulose fiber	Loose fill	3.7	The most common loose fill insulation sold. Basically, it is shredded paper, and is flammable and not moisture-resistant. Therefore, buy only those brands that have been treated with a fire retardant. Install with a separate vapor barrier.
	Blankets	4.0	
	Batts	4.0	
Urea formaldehyde	Foam-in-place	5.0	Should be installed only by a qualified and reputable contractor. Moisture-retentive; fire-resistant. Formaldehyde gas, released during curing process, may be harmful.
Urethane	Rigid boards	6.2	Excellent insulation but highly flammable, emitting cyanide gas when burning. Foam should be installed only by a qualified and reputable contractor.
	Foam-in-place		
Isocyanurate	Rigid boards	9.0	Most effective insulation available, but very expensive and, like urethane, very flammable, emitting cyanide gas when burning. Moisture-resistant.

Where to Insulate

There are two cardinal rules for insulating a house:

Surround the space you occupy and heat; and **start at the top.** Let's examine these rules.

■ **Surround the space you occupy and heat:** If you think of your house as a more or less permeable envelope containing the space you live in, then the objective of insulating becomes clear: to make that envelope *less* permeable and *more* completely contained.

Ideally, you should insulate every surface between your house and the outside world, including any surface that meets an unheated or uncooled portion

Where to Insulate and in Which Order

of the indoors on one side and your living space on the other. These surfaces define the boundaries of your energy flow, and therefore require your attention if you are to reduce energy leaks. These boundaries include every exterior wall (including the walls that connect different portions of a split-level building); and every wall, floor, or ceiling between your living area and an unheated portion of the indoors (this includes the porch, garage, attic, basement, utility room, crawlspace, cellar, and overhung toolshed, as well as unheated attic floors, finished basement walls, finished attic ceilings and knee walls, and dormer ceilings and walls).

■ **Start at the top:** Since heat rises, it stands to reason that it will leak out of the top of your house first. And because the summer sun beats down from overhead, heat also enters your house from the top, at exactly the time of year you don't want it. Therefore, the place to start insulating your home is the roof above your highest heated room, or the floor of your unfinished attic. From there, work your way down to exterior walls, and finally to floors. Give priority to floors that are, in fact, exterior lower walls (such as the floor of a cantilevered room above a carport), and to floors that have drafts or wind blowing under them.

The drawing below indicates where to insulate a house, and in what order you should do so.

Bear in mind two things: First, you don't have to do all your insulating at once. You can take it a ceiling, wall, or floor at a time. You don't even have to do it all this year.

Second, remember the law of diminishing returns: At some point, the money you spend on insulation is no longer being recouped in energy savings, and you're making a $100 improvement for a $1 per year return on your investment.

So before you run out and buy a few acres of fiberglass batts, see how much insulation you already have. Then calculate the amount of additional insulation that will bring you the best return when placed in the part of your house that needs it most. If you have any insulation at all in one strategic area and none in another, you'll probably do better to add *some* insulation to the place where you have none than to put in *additional* insulation.

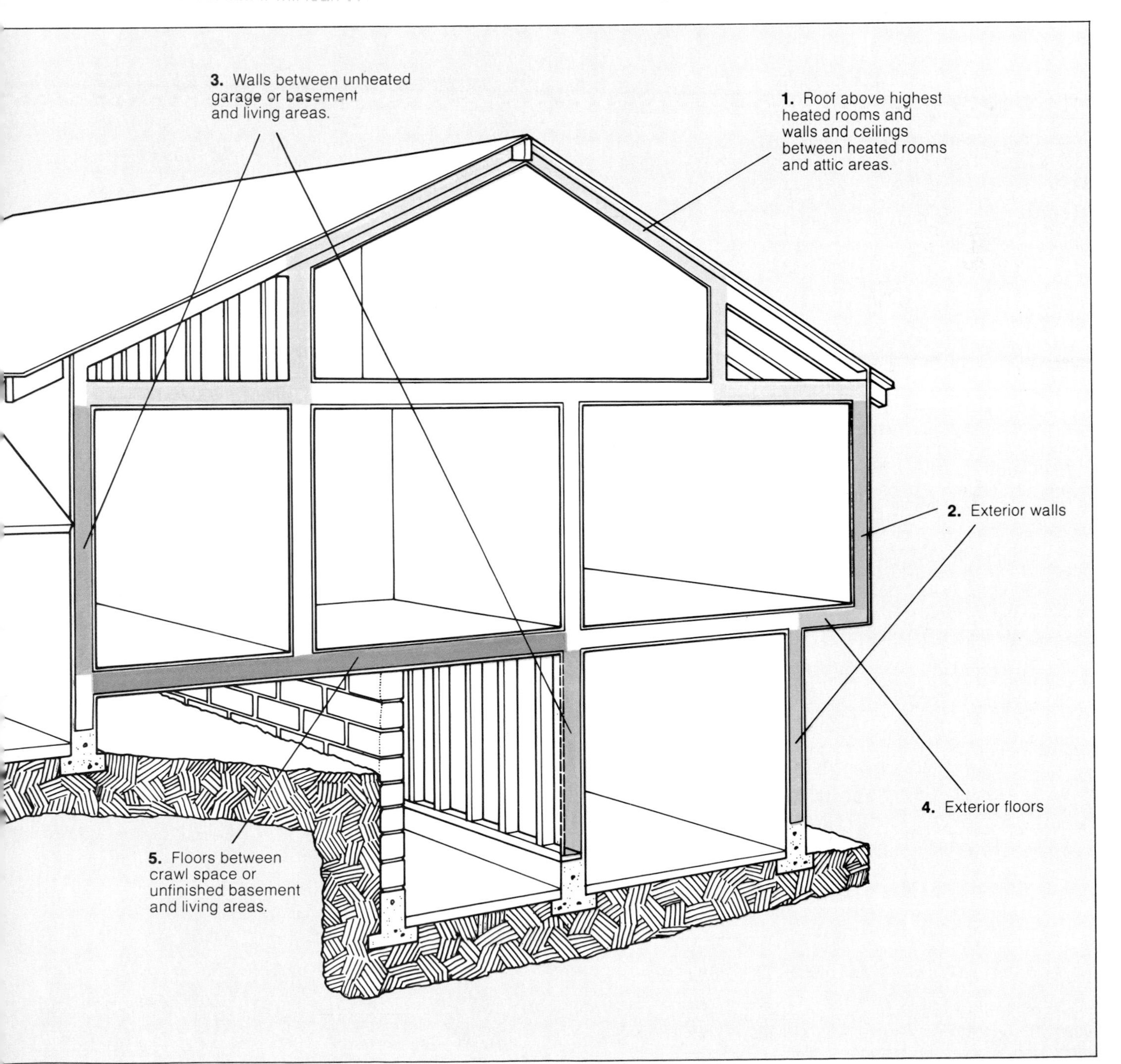

How Much Insulation Do You Need?

The next question, then, is how much insulation do you need? And the answer—like so many answers about energy conservation—is: That depends.

Like the general field of energy development, energy conservation is relatively new as a major set of government and consumer programs. Everyone investigating, studying, or writing on the subject has some powerful opinions about the potency of the insulation you should have.

How much is enough for *you* depends on both personal and general factors. The personal factors include: what temperature makes you comfortable; how long you plan to live in your house; the age and construction of your house; the number of people who live in your house; and how much time those people spend indoors, and at what activities. The more general factors include the climate in your part of the world, and the climate on your block—which may be quite different from the climate just around the corner. Some energy reports have urged people living around Detroit, Michigan, for instance, to insulate their attics to R-30 and even as high as R-33. Others have found that the point of diminishing returns is reached at R-19—that is, further insulation does help, but it is more expensive to install than the energy it will save.

After the first few inches of insulation (for example, a blanket or batt), additional insulation may be far less cost-effective. In most parts of the Midwest, for example, it will take about ten years to gain sufficient energy savings to equal the expense of adding ceiling insulation from R-11 to R-19. However, when you upgrade your insulation from nothing (about R-3 in most houses) to R-19, it will almost certainly pay you back in about five years. For a savings-estimate formula, see page 17.

■ **How much do you have?** When you did the audit, you discovered *how much* and *what kind of* insulation you already have in different parts of your house—ceilings, floors, and walls. For a closer, more specific look, here are a few easy ways to test your insulation almost anyplace indoors.

If the wall, floor, or ceiling in question is covered, you could pry off a small piece of the covering. However, if it looks like prying off a piece might be damaging or messy, drill a small hole through it, in some dark corner. Then insert a pencil, length of dowel, or other thin, straight object until you feel resistance. That is where the insulation begins. Mark your stick at that point, and push it through the insulation until you reach the next point of resistance, which will be either the vapor barrier or the far wall. Mark your stick again. The distance between the two marks is the thickness of your existing insulation. If you can hook a bit of it on the end of your stick, you can find out what kind of insulation it is, as well.

Another way to identify your insulation is to go through an electric socket. *Turn off your electricity first,* or your discovery could prove to be a rude shock. Lighting your way by flashlight, remove the light switch or socket cover and look around the in-wall installation. If you see a vapor barrier, you almost surely have blankets or batts. If you find insulation but no vapor barrier, you probably have fill or foam-in-place. Try to hook a wisp of it with a wire coat hanger, for identification purposes. If you don't recognize it by the illustrations on page 39, show it to the salesperson at your hardware store or building supply outlet. Loose fill may have settled; if so, in winter the wall will be appreciably cooler toward the ceiling than toward the floor.

You can also go through the TV an-

How Much Insulation Do You Have?

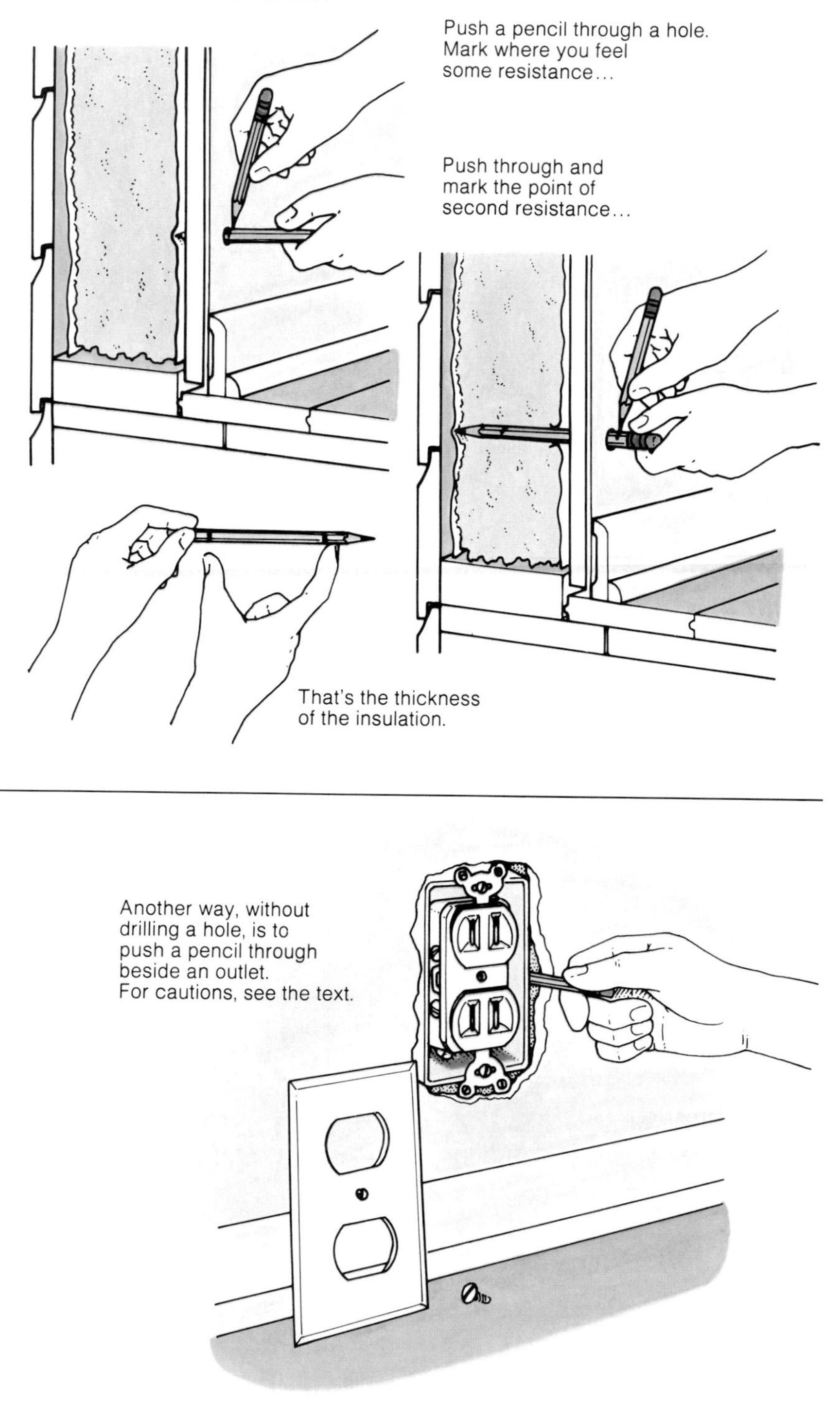

U.S. Heating Zones*

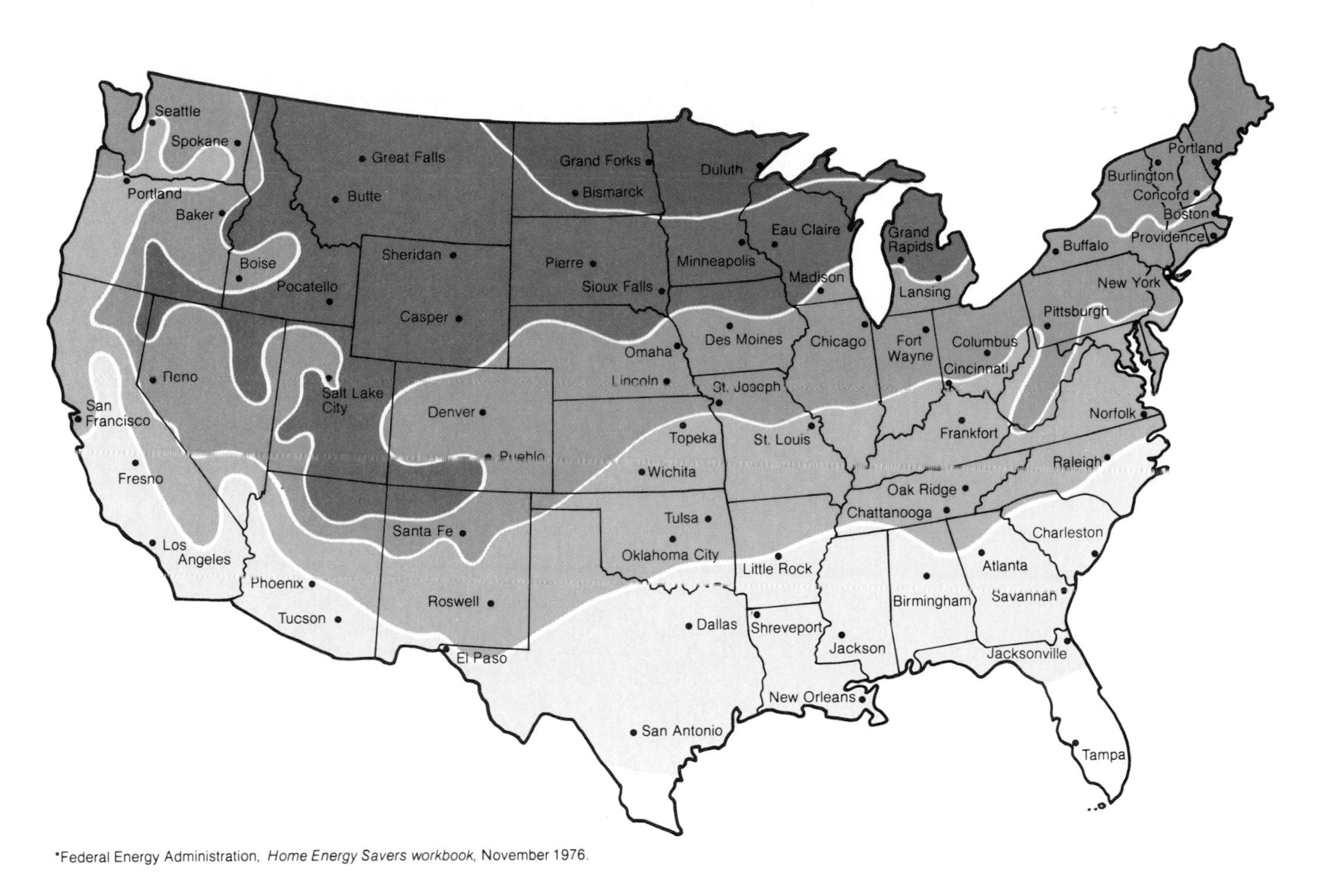

*Federal Energy Administration, *Home Energy Savers workbook,* November 1976.

Zone 1	Zone 2	Zone 3	Zone 4	Zone 5

tenna, the cable TV outlet, and dead wall-plates covering possible future socket holes. This will keep you from having to deal with live electric outlets.

When you know how much of what kind of insulation you have in which parts of your house, you can determine what additional insulation you want to install. In adding your insulation, make certain you do not cover those vents that will keep your attic ventilated (see Ventilation, pages 58-59).

As we've indicated, how much insulation you actually need depends on your immediate climate—your location with regard to sun, wind, large trees, and so on. We'll discuss the nature and use of specific climates at greater length in our chapter about the basics of solar energy. For now, the map and charts on this page will provide you with some general guidelines.

■ **Estimating how much insulation to buy:** Before you start, you should know not only how deep you want your insulation to be, but also how many square feet of space it must cover. Make this calculation by multiplying the length of your attic floor by its width. For an "average" uninsulated attic of 1400 square feet, you should allow about a day or so to accomplish the work, once you've got your materials on hand.

Recommended R-values

Heating Zone	Attic Floors	Ceilings Over Unheated Crawlspace or Basement	Exterior Walls
1	R-26	R-11	R-value of full wall insulation, which is 3½" thick, will depend on material used. Range is R-11 to R-13.
2	R-26	R-13	
3	R-30	R-19	
4	R-33	R-22	
5	R-38	R-22	

Inch Equivalents of Various Insulations for Different R-values

Insulation	R-11	R-19	R-22	R-30	R-38
Fiberglass batts/blankets	4"	6"	6.5"	10"	12.5"
Urethane foam-in-place	2"	3"	3+"	4.5"	6"
Rock wool batts/blankets	3"	5.5"	6"	9"	10.5"
Fiberglass loose fill	5"	8.5"	10"	13.5"	17.5"
Rock wool loose fill	4"	6.5"	7.5"	10.5"	13.5"
Cellulose loose fill	3"	5"	6"	8"	10.5"

How To Insulate

Once you know where to add insulation to your house and how much to add, you can begin to actually do it. The following step-by-step instructions are for adding blanket insulation, with a vapor barrier attached, to an unfinished attic floor. This procedure will show you what's involved in most insulation installations. We chose an unfinished attic because that's the very first place to insulate a house that's starting from ground zero—no insulation at all.

First, make sure to repair all leaks in the roof—there's no point inviting water damage to your new insulation.

To begin the insulation installation, take the following materials with you to the attic.

1. Insulation;

2. One or more protected, movable hanging lights;

3. Some wide boards to lay across the ceiling joists so you can kneel and stand while working;

4. A pair of heavy-duty scissors, and a matt knife, linoleum knife, long serrated knife, or saw, to cut the insulation;

5. A heavy-duty staple gun, with ample staples, and a hammer and tacks;

6. Sturdy work gloves;

7. A respirator and goggles (not mandatory but highly recommended), and a long-sleeved shirt to wear—you want to minimize direct skin contact with the insulating materials.

When you walk on your temporary board-floor, keep your weight over the ceiling joists rather than over the empty spaces, in case the boards aren't as strong as you thought.

Do not unwrap the insulation until you are ready to begin laying it in place.

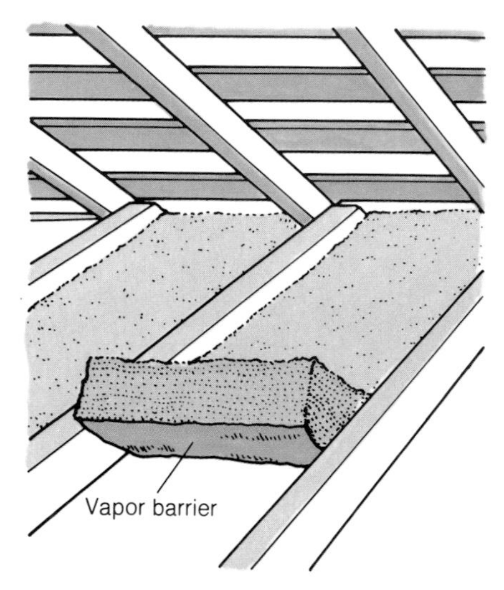

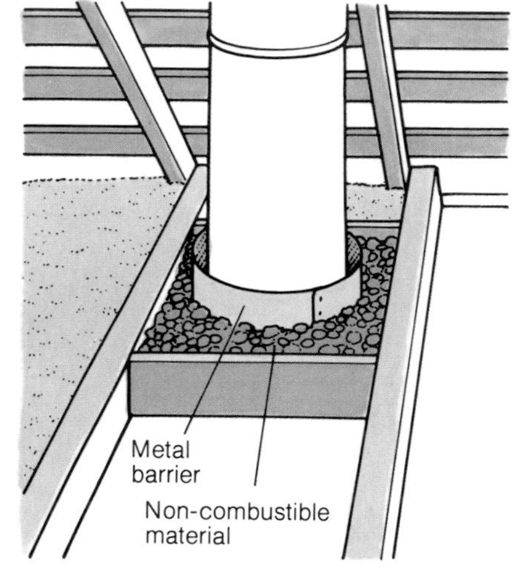

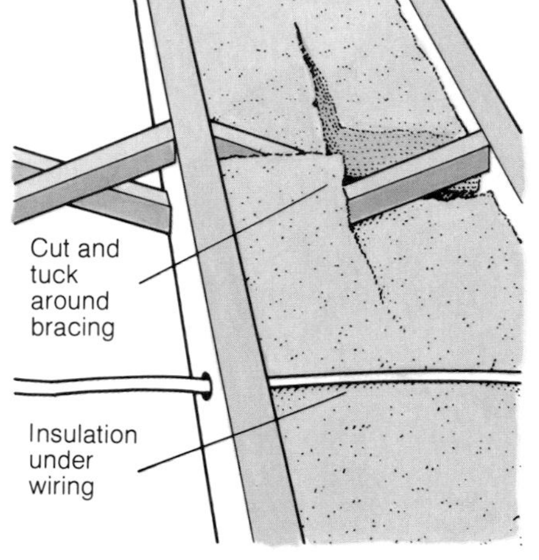

1. Beginning at the eaves to one side of the attic, place the blankets between the joists with the vapor barrier down, *toward* the living space you want to insulate. Work toward the center, pressing the insulation into the joists firmly but gently so that you don't compress the material.

2. Provide a nonflammable buffer between the insulation and any chimneys. Wrap the chimney with a metal barrier, from the bottom of the floor to 2 inches above the joists, and then surround it with noncombustible material.

3. Place insulation *under* wiring. (If you have knob and tube wiring, be sure to cut the insulation to fit down next to the wiring—not over or under it. Also, keep aluminum vapor barriers away from the wiring so you won't cause short circuits).

4. Separate the layers of insulation to surround cross braces. A short, straight cut up the center of the batt or blanket will allow you to tuck the insulation tightly around the joist braces or pipes.

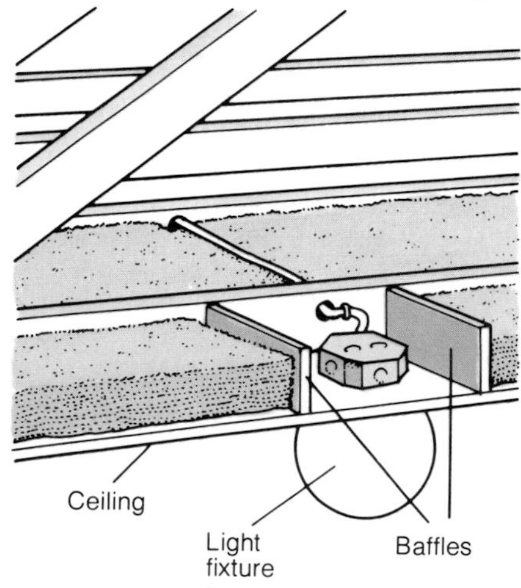

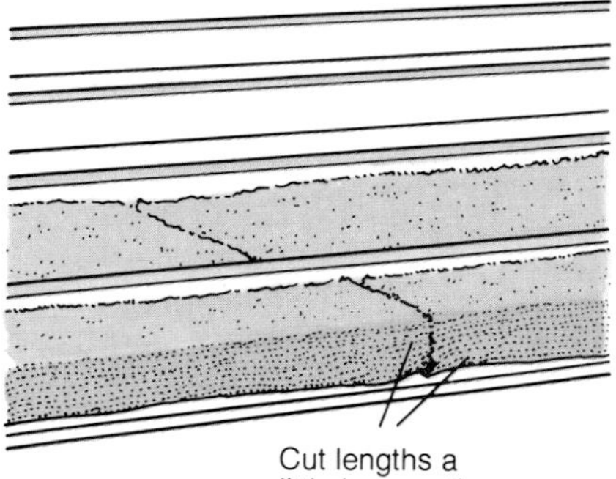

5. Do *not* block vents or ducts. Place a baffle between the vent or duct and the insulation.

6. Keep insulation at least 3 inches away from any lighting fixture or heating vent. Not only is this a requirement of the National Electrical Code, but it is also good sense. Light fixtures are electric heaters. If you insulate a 100 watt fixture, heat can build up to the point of starting wood to smoulder. This is doubly dangerous if the insulation itself is flammable. You can install a piece of sheet metal or a baffle around any fixture, outlet, or vent as a means of ensuring that the insulation stays out of the way.

7. When you near the center of the attic, start over again from the eaves on the other side. Compress the ends of the insulation together where they meet in the center of the room to ensure a tight fit.

Cutting Blankets or Batts

If you have to cut a batt or blanket, place it on a wide piece of plywood or other hard working-surface. Using one of the temporary flooring boards to compress the insulation at the cutting line and also as a straight-edge for cutting, cut the insulation with your matt knife. If the matt knife is not long enough any strong, long, sharp, serrated kitchen knife or scissors will do.

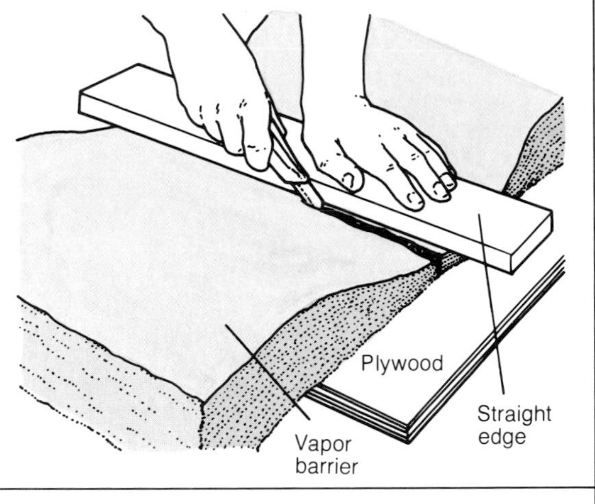

Insulating Under Floors

When you're installing insulation under floors to protect the living space above, make sure the vapor barrier faces *up*, toward the space you want to insulate. You can hold the insulation in place with chicken wire or a home-made system of wire netting, nailed to the joists.

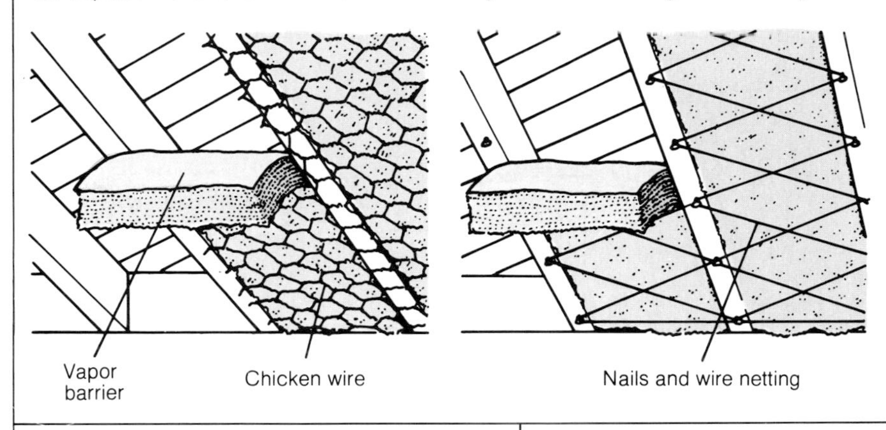

Insulating an Unfinished Wall

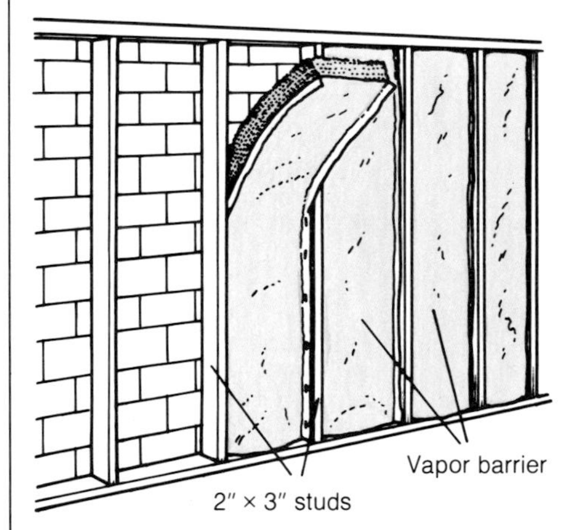

When adding insulation to an unfinished wall in an area such as a basement or building foundation, you may first have to build in a series of 1 by 3 inch studs. Next, add batts or rigid board insulation. If the insulation has no vapor barrier of its own, cover the insulated wall with heavy-duty foil or 2 mil polyethylene sheeting, stapling the vapor barrier into the studs, and repair any gashes with waterproof insulating tape. If your insulation is flammable, make sure to use a fire-retardant wallboard when finishing off the wall.

Adding to Insulation

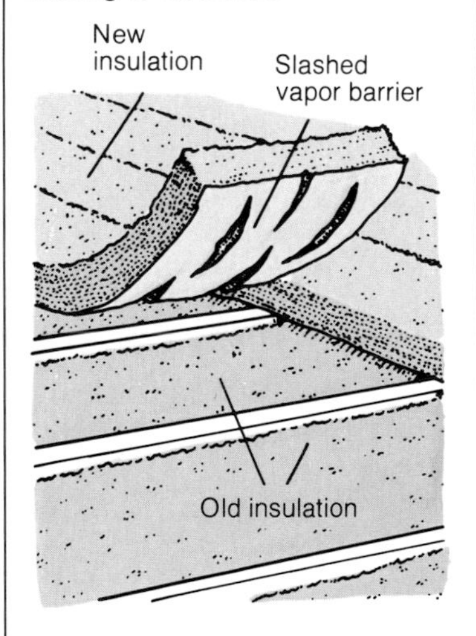

When adding to preexisting insulation, it is important to keep moisture from accumulating within and between the separate layers. If possible, use batts or blankets without a vapor barrier for your addition. If your new insulation *does* have a vapor barrier, slash the foil liberally. Then set the insulation in place perpendicular to the old insulation, and with the slashed foil facing it.

Insulating with Loose Fill

When installing loose fill, first put down a vapor barrier.

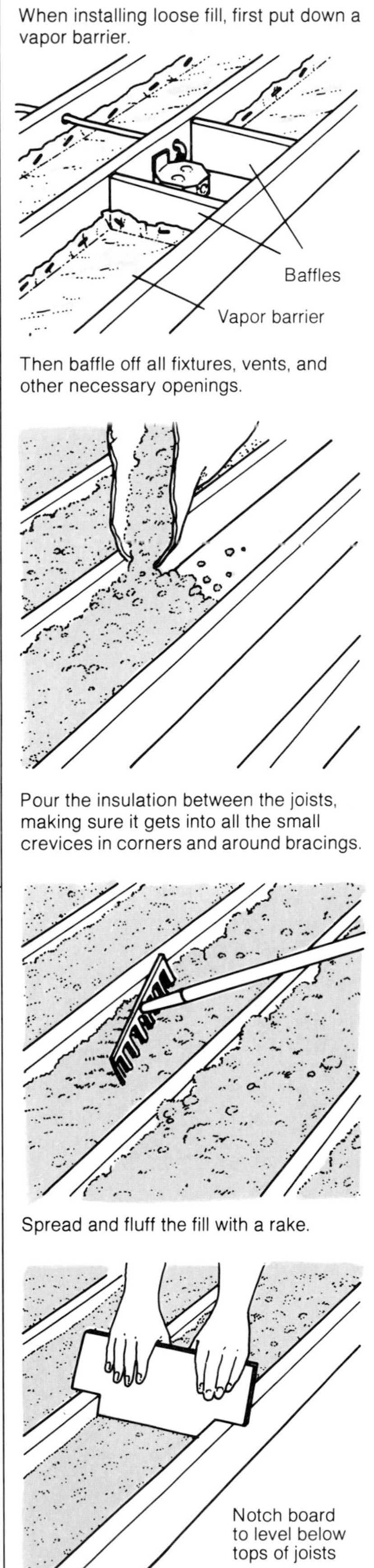

Then baffle off all fixtures, vents, and other necessary openings.

Pour the insulation between the joists, making sure it gets into all the small crevices in corners and around bracings.

Spread and fluff the fill with a rake.

If it is deep enough, you can level it off with the edge of a board.

Landscaping for Energy Conservation

Landscaping for energy conservation is not a new idea. Before the age of cheap heating fuels, it was common knowledge that a shelter should be built to take advantage of prevailing winds, the angle of the sun, and topographical features such as hills, lakes, and stands of trees. Before the invention of air conditioning and electric fans, large trees were often planted around houses to provide cooling shade in summer. When the only heat a family had in winter came from the wood they cut, windbreaks were used as an important method of conserving heat.

During the few short decades between the development of more advanced energy sources and their rising costs today, little thought was given to most of the once-commonplace energy-saving considerations. But now that energy conservation is again important, we can learn from the past and adapt that knowledge to our current needs.

■ **Microclimates:** If you've ever climbed or walked around in the mountains at 10 or 12 thousand feet on a blazing hot summer afternoon, you may have noticed a few small patches of snow nestled into north-facing crevices of thick granite walls. These cold, white anomalies exist in tiny microclimates that are related to, but quite distinct from, the larger climate that surrounds them.

The same kinds of climatic discrepancies exist around your own house, although probably to a more moderate degree.

Your geographic region has a climate you might describe as "cool," "temperate," "hot and dry," or "hot and humid." Within that regional climate are numerous local climates; in a hot, dry region, a stream may run on a green valley floor that gets little sun and is really rather cool and moist. And the local climate itself may be made up of multiple microclimates.

You can take advantage of your regional, local, and microclimates to save energy in all sorts of ways that will also make your house more comfortable and attractive. In assessing your needs and applying your solutions, bear in mind that in some ways your particular microclimate may differ even from that of your next-door neighbor. Always look at your *own* situation closely.

■ **Working with your landscape:** Depending on your location, the natural landscape may help to conserve energy. Plants at ground level evaporate moisture, cooling the air all summer. A small body of water, set between the southerly wind and the house, will provide natural air conditioning in warm weather.

There are also structural changes you can make to your house that can be a very effective means of cutting energy bills. Awnings and overhangs are two valuable additions, they are covered on page 35.

When you're looking for ways to conserve energy and cut down on heating and cooling bills, it's worth considering all aspects of your property. There may be unexpected savings that only a landscaping project could offer.

■ **Windbreaks:** While wind directions vary, they are fairly predictable within any general region. In most parts of the United States, except along the coasts, cold winds rush down from the north in winter and soft breezes waft in from the south in summer. From the viewpoints of both comfort and conservation, it makes sense to encourage the soft breezes and discourage the cold winds.

A **windbreak** is any sort of barrier that stops the wind. It can be set up on the north side of your house and, depending on the intensity of the winds, it may reduce your winter heating load as much as 20 to 30 percent. A windbreak should be constructed to allow *some* wind to pass through it; otherwise, the uneven speeds of passing winds may create eddy currents, which in turn may create unexpected winds on the far side of the house.

Windbreaks can be constructed of masonry or wood, but the most effective ones are natural: trees, tall shrubs, even networks of vines. The greenery not only acts as a block and drag on the wind, but it also gives moisture to the air that reaches the house, which raises the comfort level for the residents in cold weather. A thick screen of vegetation can also absorb some noise, which could be another consideration if you live in a city or near a highway.

Should you consider planting or building a windbreak? There's no universal answer; but if you have already weatherized your house as completely as possible with caulking, stripping, insulation, storm windows, and so on, and if nevertheless there's enough infiltration to wiggle a sheet of cellophane hung by a window on a wire hanger, the answer probably is *yes*. On the other hand, you may want a windbreak long before you know the effectiveness of your other weatherization measures.

Unfortunately, you can't measure the quantity or quality of wind in order to determine whether or not to have a windbreak. There is no R-value for windbreaks, and deciding to plant or build one is largely a subjective matter.

If you do put up some sort of windbreak, the most logical location is between your house and the direction from which the prevailing cold winds blow. As to size, there is a traditional rule of thumb: choose a tree that will reach 1½ times the height of your house at maturity, and plant it at a distance equalling about five times the height of your house. (Of course, if you are building your windbreak rather than planting it, you needn't wait for it to grow to the right height.) But this rule of thumb is only an approximation. If you live on the windy side of a hill, your windbreak may have to be higher or closer than

Landscaping for Conservation

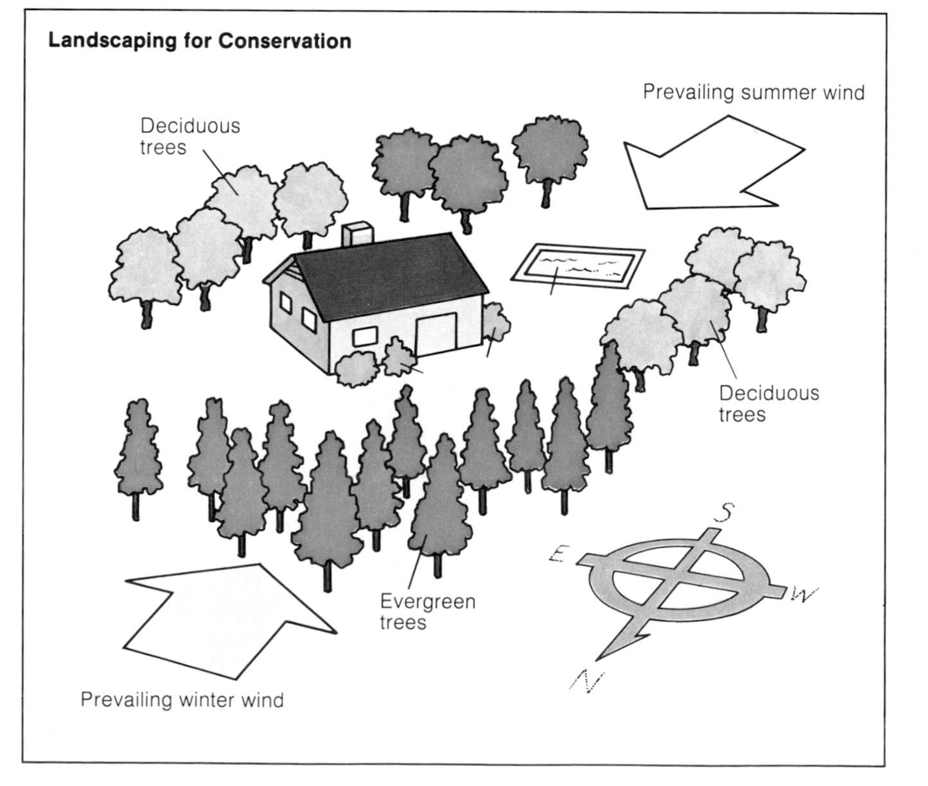

tradition dictates. A broad exposure to a wide field of wind may require a longer break than a narrow exposure will. And the limits of your property line or other encumbrances may prohibit you from following the ideal guidelines.

The same problems that apply to any subjective matter also bedevil the question of windbreaks. You must assess your own situation; modify that assessment according to your own personal needs, desires, and abilities; and choose the likeliest solution. Our guidelines should help you arrive at the best compromise possible.

■ **Trees:** There are basically two kinds of trees: evergreen and deciduous. Be-

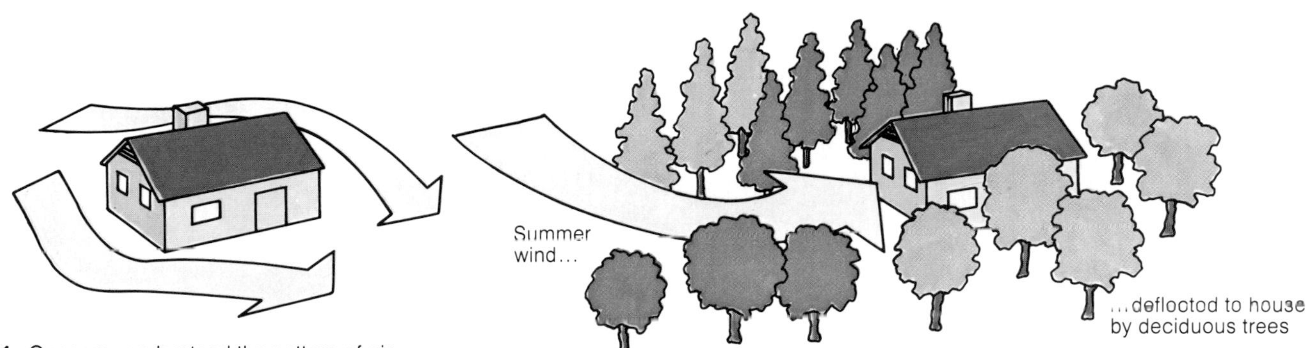

1. Once you understand the pattern of air movement around your house . . .

2. you can use trees and other landscaping features to direct warm summer breezes into the house,

3. while channeling cold winter winds away from it.

4. Closely planted, dense-growing shrubbery can be used to block wind at the same time it provides an insulating dead-air space next to the house.

5. In general, low plantings admit the sun, while tall ones block it out.

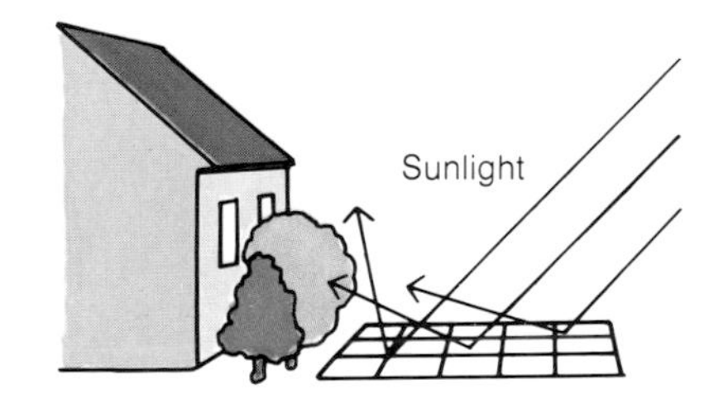

6 But low plantings can be used to block *reflected* sunlight.

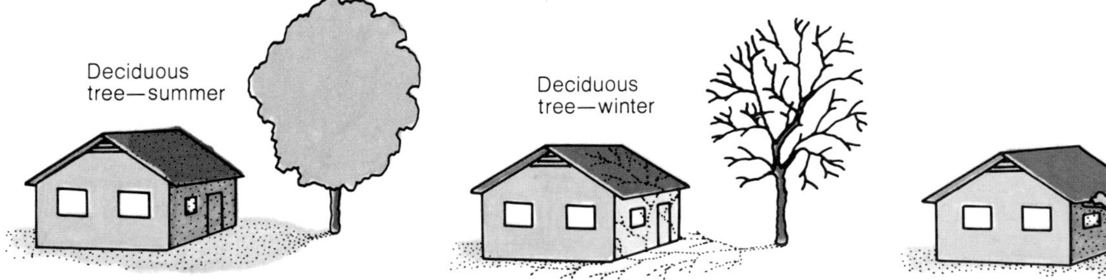

7. Tall deciduous trees, planted on the south side of your house, will block the sunlight during the hot summer months, and admit it during the cold winter months. Similar plantings on the west side of your house will help to block afternoon sun in summer as well.

8. Deciduous arbors and trellises can have much the same effect.

cause evergreens retain their foliage all year, they are particularly useful on the north side of a house, where they can provide the greatest shelter when the cold winds blow. On the other hand, because deciduous trees lose their leaves in the fall, they are excellent protection for the west, east, and south sides of a house. Their broad foliage helps reduce the heat all summer, and then lets the sun shine through in winter, when the leaves are gone. Deciduous shade trees have been shown to make a difference of 20°F to 40°F. in the attic temperatures on a warm, sunny day. Small trees can reduce reflected heat from pavement near the house, reducing indoor temperatures in the summer.

Trees used expressly as windbreaks should be positioned as noted above and should be capable of maturing to the proper height for maximum effectiveness. The number of trees and the distance between them will depend on the breadth of their root systems and the eventual span of their branches.

As you might imagine, where one tree breaks a little wind, a stand of trees can break substantially more. If you've ever flown over the Great Plains on a clear day, you've looked down on vast farms whose building complexes are tucked off in a corner by themselves, protected by several rows of trees that have been planted to form a wedge that faces into the prevailing winds. The farmhouse itself is nearly always in a pocket of this wedge, and shaded on the south by tall deciduous trees.

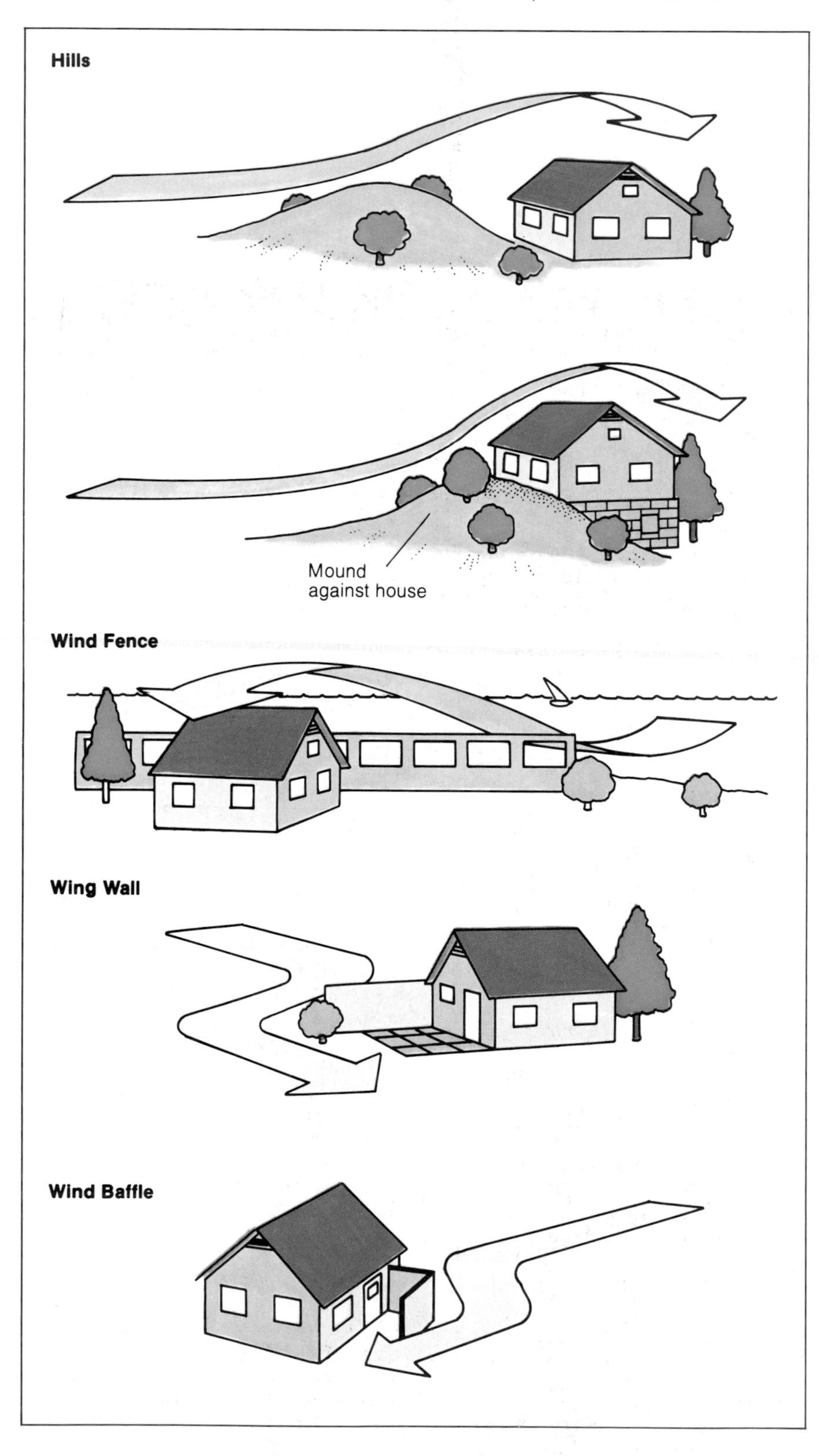

■ **Shrubs:** Tall shrubs, planted densely and close to the house, offer a layer of natural protection that cuts down the wind's impact on the building. An experiment in South Dakota compared two homes: one with a shrub windbreak, and one without. The home protected by shrubs consumed 25% less energy. Shrubs also provide an insulating dead-air space. Here, the dead-air space works on the same principle as it does in a storm window: since static air transfers heat poorly, the space where the air is trapped (in this case, between the foliage and the siding) *becomes* insulation. For this reason, shrubs are preferred over trees for plantings near the house because of their more compact nature. Shrubs can also reduce heat reflected from pavement near the house.

■ **Vines:** Vines can help cool your house in summer and warm it in winter. A variety of vines can be trained to trellises immediately beside the house, or grown to climb a bare south or west wall to shield it from summer sun. Evergreen vines can be grown in the same way on a north wall to provide some winter insulation. A lathe shade structure covered with a deciduous vine can be particularly effective on the south. You can create a green overhang, an awning or a ceiling over a patio. Each will reduce both direct and reflected heat, performing much the same function as a deciduous tree. Ortho's book, *Garden Construction Know-How* contains detailed instructions for building such shade structures.

■ **Hills:** Hills, whether natural or man-made, can offer protection against the cold. A man-made hill—a *berm*—works effectively and immediately. A berm can direct the movement of cold air around the house, encouraging it to run off down a slope instead of catching next to the foundation like a pool of water. If the berm is high enough, it can even deflect the wind over the house. Perhaps its most effective use is as insulation. Consider mounding soil up against your house to take advantage of its natural insulative qualities, but *only* if your home has the appropriate masonry structure to accommodate the constant moisture sitting against it. Because a berm can be planted, it can blend attractively with your other surroundings.

■ **Structures:** Air flows much as liquids do. And, as with liquids, there is more than one way to control the flow and direction of air.

So far we have been discussing windbreaks. A windbreak may be likened to a breakwater pier that protrudes out from shore into a large body of water such as an ocean or a large lake, and whose purpose is to break up and dissipate the force of strong waves, especially during storms.

A **wind fence** is a different matter, and may be likened to a seawall, which does not break up the force of strong waves, but rather receives their impact directly, absorbing it and protecting the land and structures that lie behind it.

Whereas a windbreak lessens the force of wind for a defined area, a wind fence is intended to create a completely calm space immediately behind it. A wind fence may be as simple as a piece of canvas strung up in the face of the wind, or as determined as a brick wall. The former should have small holes in it to allow some wind to pass through, or it will be torn down in a strong wind; the latter should be strong enough to withstand far greater impact. A windfence can be effective if, for instance, you have strong winds gusting over your patio and into your back door. It can make both outdoor and indoor living more comfortable.

A **wing wall** is an effective way to divert most winds from around part of your house, and may create a kind of sheltered patio on which, as you sit and sip your morning coffee, your newspaper will barely be ruffled, even though nearby tree branches may sway and dead leaves may leap like lizards across the open spaces of your yard. To be most useful, the wall should be adjacent to an outside door, and should block winds from the north and west of your house. (In general, these are the ones that blow both hardest and coldest.) Again, if you have a particularly windy area near an entrance to your house, a wing wall can be very effective in reducing infiltration as well as making the outdoor living space more comfortable.

A **wind baffle** can increase your cold-weather comfort, particularly if you live in a climate where winter rubs the discomfort in with chill winds that whip through your parlor every time you open the door.

Essentially, a baffle stands mid-way between a windbreak and a wind fence. It is a mini wing-wall that extends from your door frame and blocks the direction of your strongest winter winds. Some baffles may extend several feet out from the door frame, then attach to another baffle at 90°, blocking the wind still further. Only you can decide how extensive your baffle should be.

What to Look For in Buying a House

Tradition, if not good sense, has given us a certain training in our shopping habits. When we go shopping for a place to live we know to look for certain things we want: a congenial neighborhood, proximity to schools, public transportation, parking, freeways, shopping, recreational facilities; and we know to look within our own, realistic budgetary range.

While we're looking for what's obvious to us, there are other aspects of a living space we ought to examine as well. Are the doors and windows weatherstripped? Are the various nooks and crannies caulked? Are there storm windows and doors? These are small and simple tasks to accomplish if you're already living somewhere. But moving has enough headaches of its own; why add more? Besides, if you're paying for a house you might as well get a complete one; and a house that is not sealed against wind and cold is not truly a complete shelter.

Ask about, or look at, the insulation in the ceiling, walls, and floors. Are *all* the exterior surfaces insulated so that a complete envelope surrounds the space you'll occupy? Or are you going to have to spend the first winter weekends laying batts across your attic?

What kind of heating system does the house have? That furnace is going to be your responsibility, so perhaps you'd like to examine it now. Has the present owner taken care of it? You can tell a great deal from whether the air filters are clean or in need of replacement. And the ducts: are they insulated? How about the water heater?

Will the windows give you good ventilation in the summer? Is there a whole-house attic fan?

Outside, is the house situated to take maximum advantage of the sun's heat in winter? In addition to good views, do the windows lend themselves to energy-efficient heating and cooling? From which direction do the cold winds blow? And from which direction the warm breezes? Do trees, shrubs, vines, and other landscaping features offer protection against the elements? Do the grounds allow for planting such protection?

These questions, and the dozens of related ones, should not form the whole basis of your decision to buy or not to buy a particular house. But for many years we Americans have neglected even to ask such questions, and have paid a price in energy for our neglect. Now, in the interests of saving your own time, money, and energy, be sure to take your answers to these questions into consideration.

What to Look for in a Contractor

For reasons of time or inclination, you may prefer to pass some household improvements on to a professional contractor. Indeed, your local building code may *require* that some of them be handled by professionals, especially if they pertain to wiring or plumbing.

When you hire a contractor you want to make certain that he or she is up to the job. Not only is your money at stake; but if your insulation is done improperly, for example, you could find yourself with walls that are soggy and rotting, as well as drafty, in just a few years.

Although there may be infinite variations on each theme, there really are only a few "rules" that should govern your choice of a contractor, and most are apparent once you see them written out. The problems arise when they are not apparent in advance.

First, always get estimates from several contractors. Compile a list from suggestions made by friends who have had jobs performed for them, and from your local utility company.

Second, ask the contractors you are considering for the names of a few of their satisfied clients. Call those people and ask about the work that was done for them. If possible, go and look at it. Then phone the Better Business Bureau or other consumer protection agency in your area and find out if there have been any serious complaints against the person or company you are considering hiring. Don't be deterred by an occasional bad reaction, but do be concerned if you hear a lot of them.

Third, get a written contract from your chosen contractor. *Every* term you and the contractor have agreed on should be in that contract; and anything you don't understand should be clarified. Any contract you sign should include a complete and specific list of the work to be performed, the materials with which it is to be performed, the dates the job will start and end, indication of warranties, and a list of liability provisions.

Most of the people who do the sorts of home maintenance jobs that have been discussed in this chapter are honest and upright folks who are highly competent in their profession. They will argue for a contract that benefits them, just as you will argue for one that benefits you; and you should be able to come to a mutually satisfactory arrangement. The purpose of going to some pains to select your contractor in the first place is to ensure that you end up with someone you can trust, and who will give fair value for money received. The advantage of having the contract is that if you run into problems your contract is all the protection you will have.

Just like your car, the mechanical systems in your home will perform better, last longer, and cost less to operate when they are clean, tuned up, and in good repair. A little effort can yield big savings in a short time.

Plugging the Energy Leaks: The Systems

Congratulations! If you've followed this book's advice so far, by now your house is pretty well sealed and you're saving from 15 to 50 percent of your heating and cooling energy. You can stop here, and enjoy lower energy bills and greater comfort—or you can go even farther.

Inside your well-sealed house are the various systems that let you control the temperatures of your air and water with considerable precision. Since you've plugged up the holes in and around your doors and windows, these systems are no longer doomed to spill their heated or cooled air outdoors. So now it's time to make sure they are doing their jobs as efficiently as possible. Making just a few small adjustments in these systems will net you some large energy-saving bonuses.

For instance, it will cost you nothing in money or effort to turn your thermostat down 5 degrees. You may not even be home a third of the time, and for another third of the time you're probably in bed under the warm blankets, so you may not even notice the difference in terms of comfort. But according to the Department of Energy, even this tiny change in your living habits will save you from 14 to 25 percent of your total heating bill, depending on which part of the country you live in. Turning your thermostat down another 3 degrees, for a total of 8 degrees, pushes your estimated savings up to 19 to 35 percent.

Savings on heating costs will always be greater in colder climates, where more heat dissipates out of even the best-sealed house. But in warmer climates, where you may have air conditioning throughout the entire house, you can save 10 percent on your cooling bill just by turning your thermostat *up* 3 degrees in warm weather.

Furnace maintenance is another important way to conserve energy. You already know that keeping your car tuned helps its ability both to perform and to conserve gasoline. This same principle holds true for any mechanical system.

Some maintenance tasks, such as changing the furnace filter, should be done several times a year. But you don't have to call in your service person every time you need a new filter. In most cases, it is a simple job you can easily do yourself. Other aspects of furnace maintenance require annual visits by a qualified professional, but a $40 annual cost can save you hundreds of dollars in the long run.

Insulation does not stop with your attic or floors. Wrapping your heating ducts with insulation, particularly when those ducts pass through unheated areas, can increase the effectiveness of your heating system, and keep your heat, and heating dollars, from seeping out through unseen "holes".

Summer cooling also presents energy-saving opportunities. How your house is vented, what type of fan you have and where it is placed, how efficiently you use your air conditioning system, and what kind of landscaping you have are all factors to consider when you are trying to cut down your fuel bills.

The importance of good air circulation and proper venting is generally underestimated and can help cut down summer utility bills as well as raise the comfort level of your home. Whole-house fans can do wonders on those hot summer nights, particularly when your house has good ventilation.

Hot-water heating generally represents 20 percent of your heating costs, and there are some simple, practical ways you can reduce waste and conserve heat.

This chapter is devoted to exploring the various options open to you to increase the effectiveness of all your home heating and cooling, while reducing costs.

How to Read Your Utility Meters

You can't know how much energy you waste until you know how much you use. One of the best ways to gauge your energy use is to do what the utility companies do—read your meter. These precise, ever-watchful yardsticks keep the gas and electric companies informed about every kilowatt and cubic foot of energy consumed in your house. They can keep you informed, too.

For the most part, you are probably keeping yourself in the dark about how much energy you use. When the utility bill rolls in, do you glance at the "amount due" column, make some silent and fruitless objection, and grumble as you write out the check? By the time you read your bill, it may have little obvious relation to your prior energy consumption. The "amount due" doesn't appear to you as the oven you forgot to turn off, or the windows you left open and the furnace you left on "high" when you went away for the weekend, or that tiny leak in the bathroom faucet that drips hot water no matter how tightly you turn the handle shut.

Utility meters ordinarily are located outside the house, or someplace else that might make them hard to see on a regular basis. This means that you have to make a point of checking on them if you're going to learn how to read them. You must make them a deliberate part of your energy awareness.

A highly-recommended exercise for every member of the family who's old enough to care about such things is to turn off everything electric in the house and stand in front of the meter, watching the dials closely. Then have someone else turn on the electric appliances, one at a time and one after another. Keep watching the meter—as the hands begin to crawl around the dials, you can actually see how your energy is used.

Electric meters calculate your kilowatt usage over a specific period of time, as **gas** meters measure the cubic feet of gas you use. Both electric and gas bills always include a starting and a stopping date; it's between those dates that you pay for your use of energy. If you want to watch your energy consumption for a particular time period, be sure to note the dates of your readings.

Electric and gas meters are read straight across, from left to right, like English. However, some of the small dials within the large frame of the meter turn clockwise, and some turn counterclockwise. Be sure to notice which is which when you take your readings: the individual dials always move from 0 to 1 to 2 and so on up to 9 and back to 0. Whenever a hand falls between two numbers, read the lower of the two. Thus, the electric meter in our illustration below reads: 54698 kilowatt hours.

If you go away and come back a day, a week, or a month later, you'll see that the hands on the meter have advanced on their dials. From this number, subtract the earlier number. The number you're left with is the number of **kilowatt hours** of electricity you used between readings, or the number of **cubic feet** of gas. A kilowatt hour will light ten 100-watt bulbs for one hour, or ten 200-watt bulbs for 30 minutes, or ten 50-watt bulbs for two hours. A therm equals (1) 29.3 kilowatt hours, under conditions of **perfect** efficiency, and (2) 100 cubic feet of natural gas, and (3) the heat/fuel that on average central-heating furnace will burn in one hour.

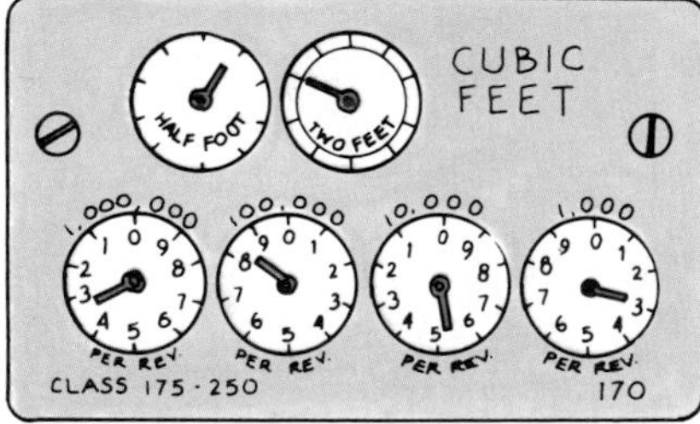

Electric and Gas Meters

Some electric meters have a direct read-out as above left. Most have dials as above. Both of these have the same reading, 54698 kilowatts.

The gas meter, left, reads 385200 cubic feet.

Furnaces

In most American homes, the primary heating system is a furnace fueled by oil, natural gas, or electricity. The furnace usually is regulated by a thermostat, which you set to the temperature you want maintained in your house. As the air cools off, it "informs" the thermostat that more heat is required. The thermostat in turn "informs" the furnace, and the furnace kicks on. As the air in your house warms up, it informs the thermostat that no more heat is needed; the thermostat passes this information on to the furnace, and heat production ceases. As the air cools down again, the cycle is repeated.

The success your furnace has in converting the potential heat in its fuel into actual and **usable** heat is called its **combustion efficiency**.

There are oil, gas, electric, and coal furnaces. If you don't already know what kind you have, take a look at it. An oil furnace has a motor, pump, and blower and has to have a fill stem somewhere for oil deliveries; and you surely know if you're getting oil deliveries! A gas furnace has no pumps or motors, but has a pilot light and gas pipes feeding in from outside. An electric furnace has many wires, but no pilot light. And a coal furnace has coal stored in a large hopper.

On a daily basis, all you need to know about most heating systems is how to turn the thermostat up and down. Even if a furnace gets worn or clogged, it may continue to work. Its efficiency, however, will be impaired—it is burning more fuel, yet heating less well. Your furnace should be checked annually.

Doing a furnace checkup may not be your idea of an entertaining home project—and in fact, for the most part, you should leave that task to a trained professional. Assuming that your furnace is not presently in need of a major overhaul (if it is, get it done without equivocation, or your living room will be very chilly come December when your furnace breaks down completely), a general tuneup will cost you about $30 to $40, and will probably save you at least the cost of the minor adjustment in unused fuel. It could even save you twice that amount.

If you don't know when your furnace was last inspected and can't find out (if, for example, the previous owners of your house moved to Tahiti and left no forwarding address), you can safely assume that it needs at least a bit of cleaning in early fall. Call in a service person to inspect it at that time, and you'll find out exactly what your furnace needs. Most furnaces are pretty solid creatures, and it is unlikely that you will have to spend much.

Furnace Maintenance

Since the combustion efficiency of **electric furnaces** is very high (almost 100 percent), there is basically no way to improve on the service yours provides except to have it checked periodically, according to the manufacturer's instructions.

Gas- and oil-fueled furnaces, however, are another story. Their combustion efficiencies rarely exceed 80 percent; frequently they drop toward 60 percent or even lower. Most of the lost heat simply slips up the flue; flue gas temperatures ordinarily fluctuate between 350° F and 700° F where they leave the furnace.

Even though maintaining your furnace is one of those tasks that's best done by an expert, it pays to be as prepared as you can—you should be able to discuss a few specific items with your service person. (For some suggestions on how to work effectively with contractors, see "What to Look for in a Contractor," page 49.)

Electric furnaces require little in the way of servicing, but at each checkup, gas and oil furnaces require both a **draft test** and a **stack temperature test**. A draft test checks the amount of air that's drawn into the furnace to mix with its burning fuel. If the draft is too great, heat is lost up the flue unnecessarily; if the draft is too small, the fuel won't burn completely. The stack temperature test determines *how much* heat your furnace is losing up the flue (all furnaces lose *some*). If your stack temperature is too high, you may need to have your furnace burner adjusted, or your heat exchanger cleaned.

■ **Oil furnace maintenance:** In addition to the tests noted above, an oil furnace should have a *smoke test* and a CO_2 test. The smoke test determines whether your furnace is burning its oil efficiently. If it isn't, there will be a heavy residue on the inside of the furnace. This residue insulates the heat exchanger surfaces and diminishes the useful heat produced by your system. The CO^2 test gauges the quantity of air mixing with your oil for combustion. Too much air will cause a larger fraction of the heat to go up the stack, rather than into the house. Too little air will produce smoke and also reduce efficiency.

At each visit, the maintenance person should change the *oil* and *air filters* and clean the *oil burner*—combustion efficiency will improve, and you may realize immediate savings. The service person should also adjust and clean the thermostat contacts in your house, and the heating elements in the furnace, if necesary. The draft regulator, oil pump, electrical connections, and the fuel/air ratio all should be checked and adjusted, if need be; oil leaks should be spotted and corrected.

■ **Gas furnace maintenance:** With a gas furnace, have the maintenance person check and adjust all valves and regulators, the air supply nozzle, and the thermostat contacts in the house. Baffle and burner housings, as well as burner bottoms, should be cleaned; and, of course, filters should be replaced (unless you plan to take on this job yourself—see page 54).

■ **Coal furnace maintenance:** Few people use coal-burning furnaces these days; and since the recent increased interest in coal production concentrates on industrial and research developments rather than on home heating, coal-burning furnaces are not likely to gain enormous popularity in the near future. However, if you already have a coal-burner, it should be cleaned and adjusted at the end of every heating season. Make sure that the stoker is cleaned and adjusted, and that the inside of the coal screw and the hopper are oiled, to keep them from rusting.

■ **Heat pump maintenance:** If you have a heat pump: change the filters every three months; always leave the temperature at one setting; and have the pump checked annually by a professional service person. Usually, you can buy a maintenance service contract from the pump's manufacturer, which will save you considerable trouble should something go wrong later on.

■ **Wood and combination furnace maintenance:** If you are one of those rare individuals who heat their homes with a wood furnace or have one of the newer combination furnaces, be sure to check with the manufacturer about an appropriate maintenance program. At least be sure all moving parts are cleaned and oiled, and replace filters as they become dirty

Heat Pump

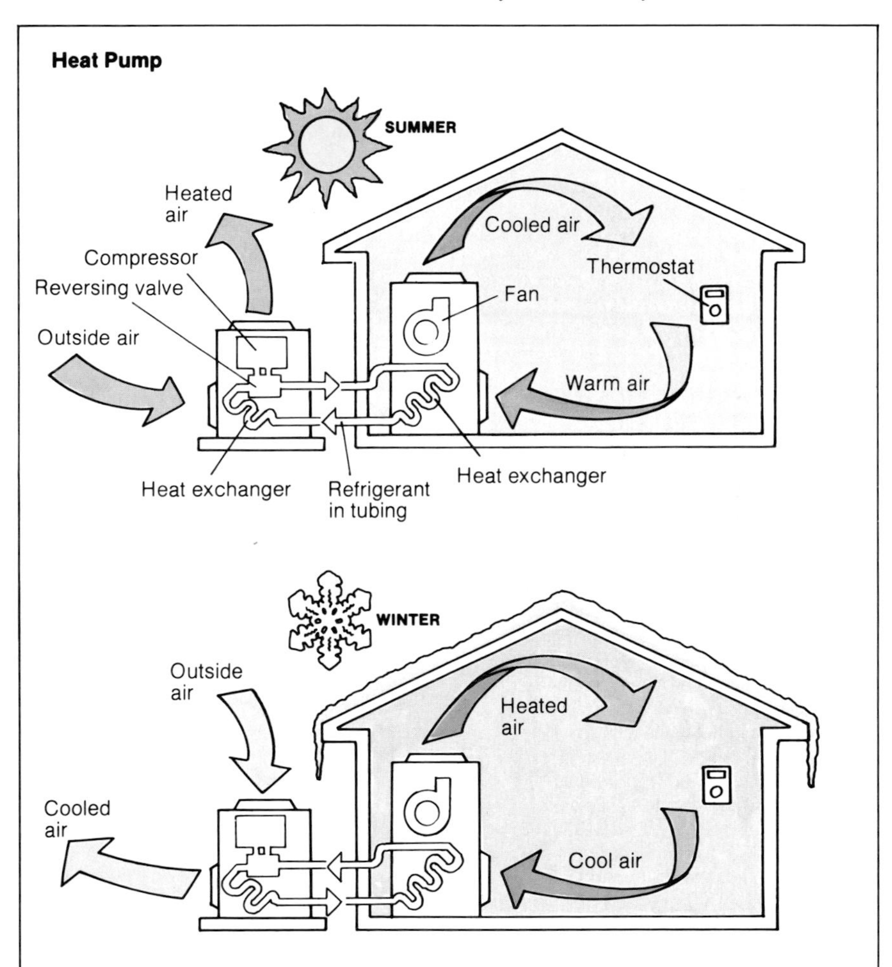

The heat pump is an automatic heating and/or cooling device, which uses a liquid refrigerant and has components similar to those of an air conditioner. Essentially, the heat pump refrigerates the outdoors and pumps the heat it picks up in this process into the house. In summer, it reverses the process. Heat pumps cost 15 to 20 percent more than conventional furnaces, but depending on where you live and the cost of fuel in your area, this device can save up to a third of your heating costs and pay for itself in as few as three years. Heat pumps tend to be more economical in warm-climate areas. If you are considering replacing your furnace, you might want to look into them. Ask your local utility company for assistance.

How a Furnace Works

■ Forced-air systems: Most forced-air furnaces work on the same principles. An outer covering (the part you see) protects an inner shell known as a *heat exchanger*. This is where fuel is burned. In gas, oil, and coal fired furnaces, the air inside the heat exchanger itself is vented outside the house through the flue pipe, so its noxious fumes never enter your living space. When the heat exchanger reaches a certain temperature (predetermined by the manufacturer), blowers suck air from the return ducts into the outer covering of the furnace. There, the air is warmed by the heat exchanger. Then the blowers blow the newly warmed air into your living space through the supply ducts.

If you have a gas furnace, ask your service person or utility company representative to show you how to relight the pilot light if it blows out. Furnaces manufactured by different companies may require different procedures. In general, though, here's what to do if your gas furnace isn't working:

1. Turn the pilot light switch to *off*.

2. Wait for any lingering gas fumes to disappear. Then turn the switch to *pilot*.

3. Hold a match to the nozzle, and push the red *reset* button. Hold the button down for the length of time required by the manufacturer—usually 30 to 60 seconds.

4. Release the reset button.

5. When the pilot light stays on, turn the switch to *on*.

If you have a gas-fired wall heating-unit, use the same procedure.

■ Pipe-heating systems: In an air system, the furnace heats air which either rises to, or is blown to, the home's living spaces through ducts. In a water or steam system, the furnace heats water in the boiler which is then pumped as water or steam through pipes to radiators or baseboard heaters. But essentially, the mechanism is the same: the furnace heats air or water which is then distributed to the desired space.

Furnace Filters

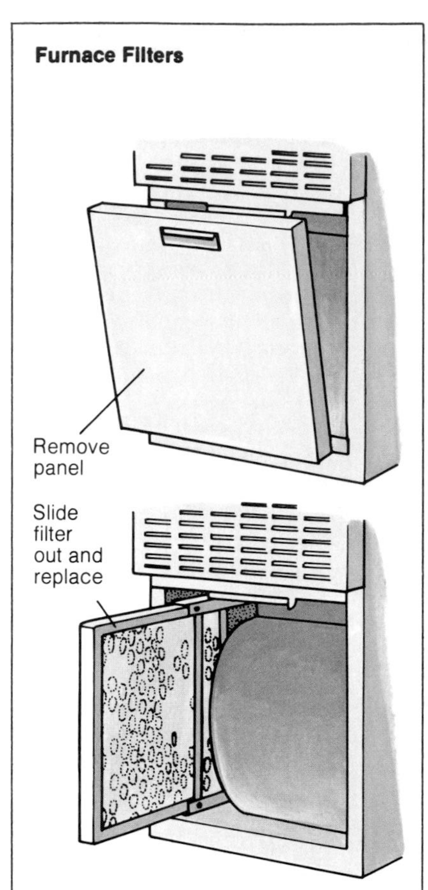

Changing a furnace air filter is an easy and inexpensive job you can do yourself. The purpose of a filter on any forced-air furnace is to keep dust, soot, and other airborne grime out of the air that blows into your living room. Air has more trouble passing through the dirt-laden filter; therefore, your furnace has to work harder when the filter gets dirty. So change your furnace's air filter *at least* a couple of times during the heating season, and as often as every month if you live in a dusty area or if your furnace has to work long hours. If your central air-conditioning unit circulates through your furnace filter, you may have to replace the filter during the cooling season as well as during the heating season. The expense of a new filter—a dollar or two, depending on the size—will quickly be made up in lower energy bills. And it will save strain on your furnace in the long run, too.

It isn't hard to change a furnace air filter. If, after reading the following paragraph, you still have questions about how to do it, ask your furnace service person to demonstrate the process during his or her next routine visit. Then you too can do the job, and with confidence.

To replace a filter:

1. Turn off your thermostat.

2. Locate the metal panel that covers the filter on the furnace, near the blower.

3. Remove the panel and slide the filter out.

4. Slide the new filter in, according to the air-flow directions marked on it. (Be sure, of course, that the new filter is the correct size for your furnace.)

General Furnace Schematic

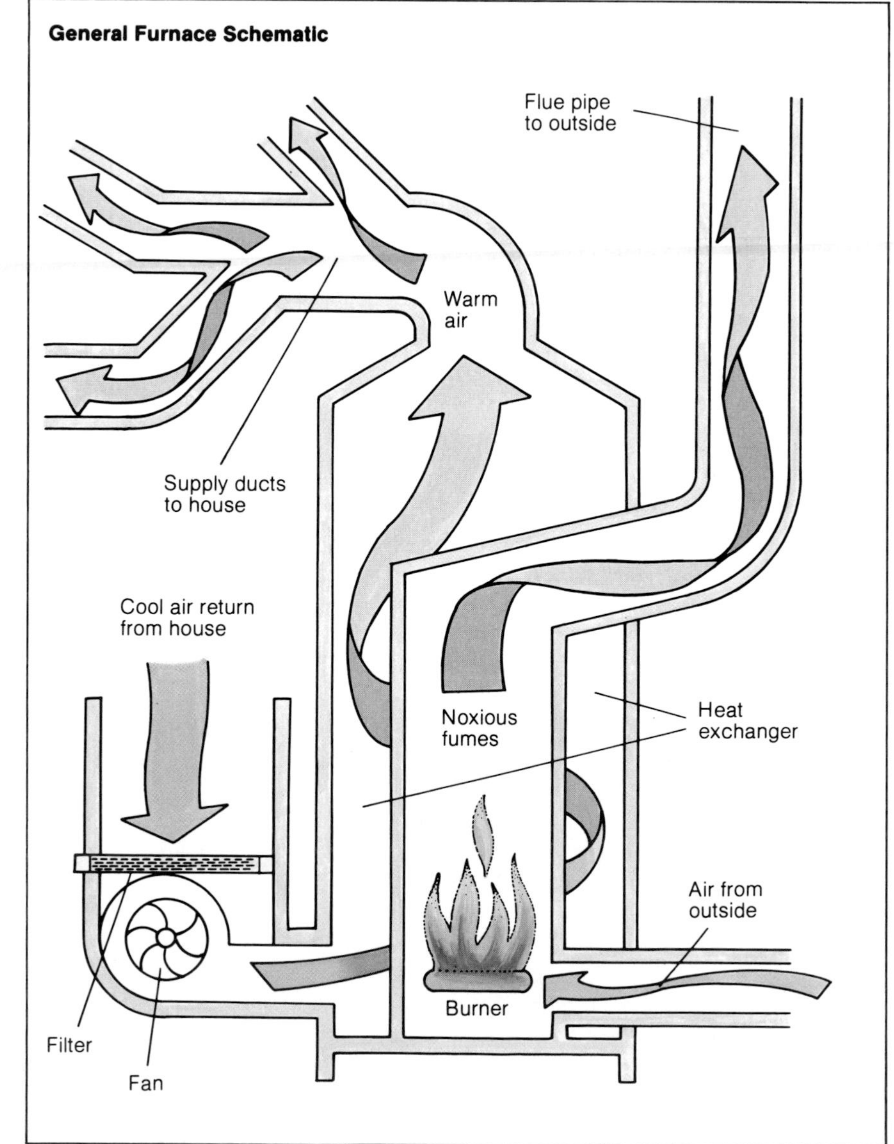

■ **Heat recovery unit:** In a properly maintained furnace, the gases that escape into the stack generally range between about 350°F and 700°F. If the temperature falls below about 300°F, the gases will leave water vapor in the flue, which can rust the vents, and may also fail to burn off some noxious components. But if the temperature rises above 700°F, a lot of useful heat is simply being wasted. Much of this heat can be recovered through the use of a heat recovery unit.

Some heat recovery units clamp onto the flue pipe; others are inserted in the flue gas duct. In either case, heat is extracted from the stack, instead of being sent up into the sky.

By itself, a heat recovery unit will not reduce the amount of fuel you burn. But it will make more of the heat your fuel generates available to you. Of course, it will be effective in inverse proportion to the efficiency of the furnace: as the stack temperature drops, so does the amount of heat that can be recovered by a heat recovery unit.

You can use the recovered heat by directing it into your heat distribution system and sending it on its way through your house. Typically, a heat recovery unit will cost $100 to $200 installed, and save about 6 to 10 percent in fuel expenditures, paying for itself in three to six years. In some areas, however, heat recovery units are not permitted by building codes, so you should be sure to find out about your local code before you buy such a device.

Radiant Heating

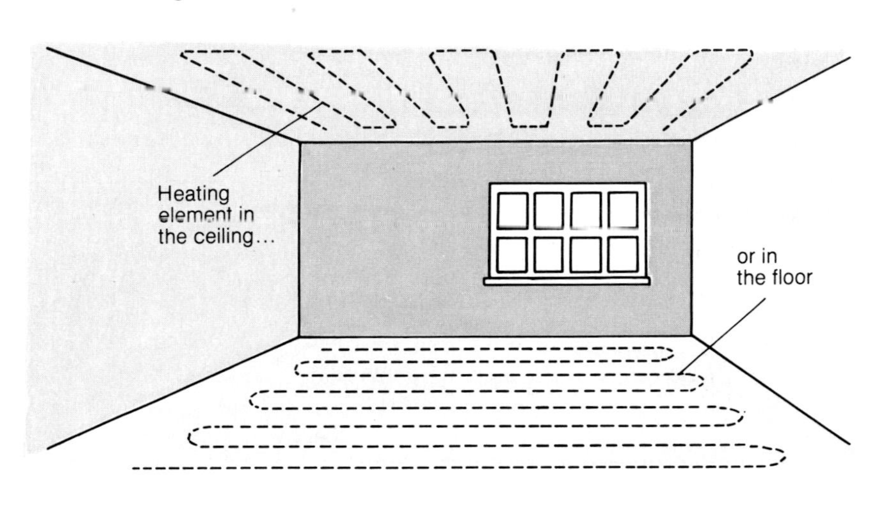

Radiant Heating

Radiant heating systems have electric cables or copper hot water pipes embedded beneath the flooring surface or (rarely) in the ceiling. Radiant heating offers certain advantages: (1) the comfort level may be quite high, even when the temperature is not. This is because radiant heat emanates directly from a large area of a room's surface; and (2) each room's heat is controlled separately, so different rooms can be maintained at different temperatures. The disadvantages of radiant heating are that: (1) the systems tend to heat up slowly; and (2) any form of electric heating is more expensive than other, more common, options.

Wall Heaters

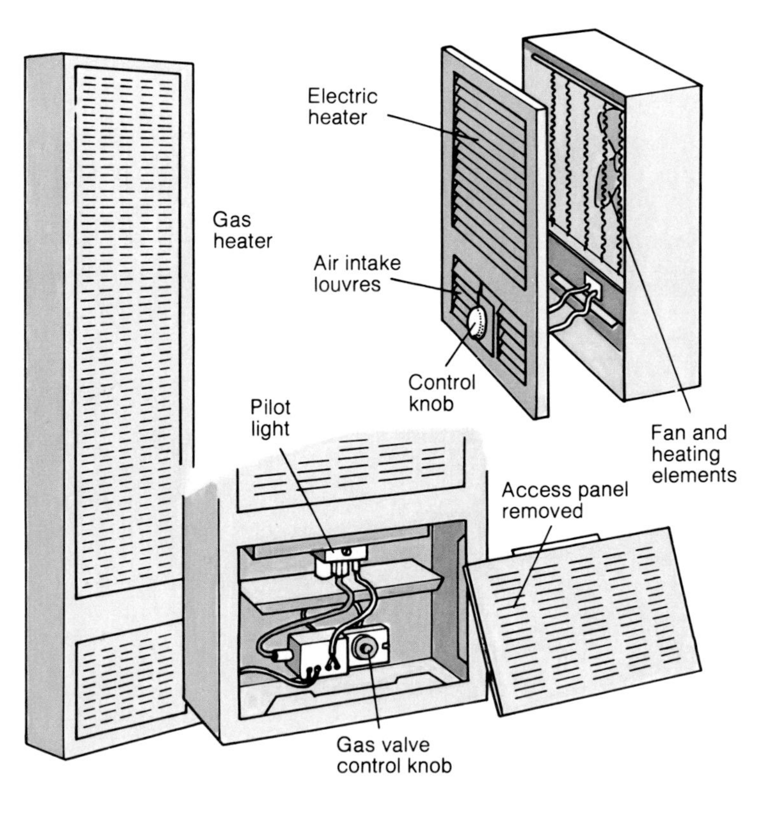

Wall Heaters

A small gas or electric forced-air heater sometimes is mounted in a wall to heat an individual room. Several such heaters may warm an entire house, if its heating requirements are low.

The gas wall heater consists of a pilot light or pilot-free spark ignition, with a gas supply on tap and a dial or thermostat, which lets you regulate the heat. When you turn up the dial, more gas gets burned sending the warmed air out through a large upper screen into your room.

The electric wall heater functions in the same way, but it uses electric resistance coils instead of natural gas to warm the air that moves across them.

The best way to maintain wall heaters is to keep them clean and free of dust. A periodic check by your utility service person will ensure efficient operation. With gas wall heaters, you can turn off the pilot light in the summer, reducing both costs and residual heat. Instructions for lighting and relighting are usually somewhere on the unit. If you're uncertain or you can't find instructions, have your utility service person do it the first time. Watch what the person is doing, and be sure you will be able to do it yourself when you are alone.

Heat Distribution

Your furnace is clean and is heating as efficiently as it should—but you still need to get that heat from the furnace to your living space.

The four principal methods of distributing heat through a house are **hot-water heating, forced-air heating, steam heating**, and **gravity convection**. You'll probably want to leave the more complicated maintenance tasks to a service person. But there are a few things that you can do to make your heating system more fuel-efficient and economical.

■ **Hot-water heating:** In a hot-water heating system, the radiators tend to fill up with air, taking the place of the heated water and reducing the amount of heat available to you. To rectify this situation, a few times a year open your radiators to let the air out. First find the valve on your radiator—it may be up near the top, or toward the bottom along one side. When water starts to pour or spurt out, close the valve. A few cautions, however: (1) do not open the valves while the heat is coming up, or you'll spray the room with hot water, steam, and hot air; (2) hold a bucket or pail under the valve (you *are* going to strike water); and (3) be careful—this water is hot!

The service person should maintain the pump and motor, as well as the flow-control valve and radiator valve; look for leaks in the pipes; and drain and flush your boiler once a year.

■ **Forced-air heating:** If you have forced-air heating, clean or replace the air filters near your furnace every month or two to keep your heating system at peak efficiency. It's worth doing yourself—it costs too much to have an expert come to your house six to twelve times a year merely to handle such a basically simple task.

At the other end of the system, take a vacuum cleaner to your hot-air registers every few weeks to ensure that they allow free passage of air.

During the service person's annual visit, make sure that he or she checks the fan blade and oils the fan bearings; and adjusts the blower operation, if necessary, checking for duct leaks and oiling the blower motor (unless it has sealed bearings).

■ **Steam heating:** If you have steam heat, drain a gallon or two of water from the bottom of your boiler at least once a month during the heating season. Otherwise, sediment will accumulate and effectively insulate your boiler from its heat source, which would be a great waste of heat, and also a source of corrosion that can eventually make the boiler leak.

Have your service person take a look at the water system in your boiler. He or she should examine the venting system for rust, and clean away dirt (especially accumulated dirt from cat or dog hairs).

Whatever kind of heating system you have, use your common sense for the details: Make sure there's adequate space around the tank so that the pilot light doesn't go out. Don't paint the radiator—that will insulate it and keep heat from reaching your living area. If you are handy with tin snips, you can make a small heat deflector from any kind of sheet metal, and attach it above your radiator or heat register to channel the heat to those parts of the room you occupy most. This sort of deflector will be especially useful if your register or radiator is located below a window, since windows—even storm windows—lose a large proportion of heat directly through the glass. You can also purchase deflectors. See page 65.

■ **Gravity convection systems:** In these systems, found in many older homes, there is no blower and, therefore, usually no filter. Heat simply rises from the furnace through ducts to the living spaces above. Since the furnace must be in the basement which is likely to be cold, it is especially important to insulate these ducts. If heat is lost through the ductwork before it gets upstairs, the furnace has to work overtime to keep the room temperature at the level you want. It is equally important to keep the vents and grates free of dust and debris so that the flow of heat is not interrupted. The vents, which are usually floor or baseboard registers, can be cleaned with a vacuum as you are cleaning house.

How Heat Is Distributed

A forced-air system blows heated air from the furnace to the living area through a system of supply ducts to room registers. The openings of these registers usually can be regulated by means of small levers to increase or decrease the amount of heat flowing into a room while the furnace is on. A separate set of ducts returns the air from the living space to the furnace.

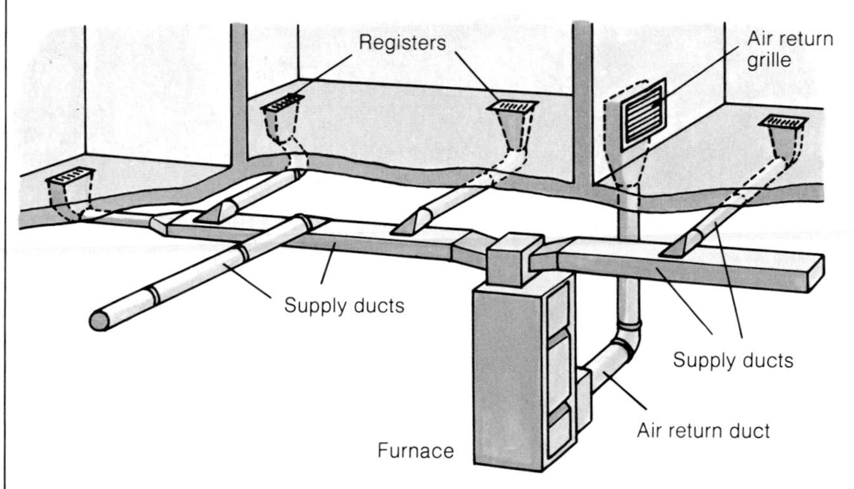

Pipe systems may use steam or hot water to circulate heat from a boiler (which is heated by the furnace) to radiators or baseboard heaters in the living area. Like forced-air systems, most hot-water pipe systems return water to be reheated through a second set of pipes. For the most part, steam which condenses to water in the radiator runs back to the boiler through the same pipe in which it got to the radiator.

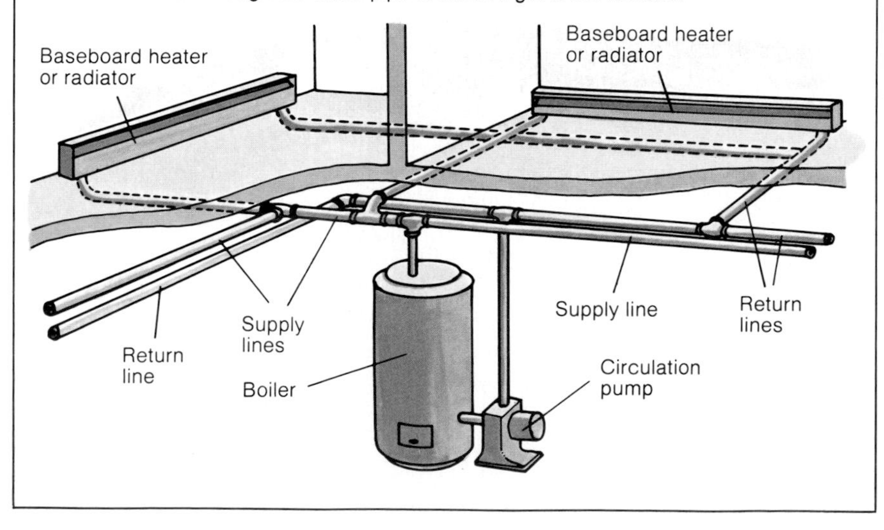

Duct Insulation

Ducts are the distribution arms of a building's heat production system. They carry warmed air from the furnace to the farthest, coldest rooms in the living space, and deliver it through vents or registers in the walls or floors. Particularly when a duct has to convey heat over a long distance, or when it passes through a garage, basement, attic, or other unheated space, much heat may be lost through holes and cracks in the ducts, and simply through the duct walls, by conduction. When ducts run through walls and floors they are pretty well beyond your reach—carving your way through a wall to get at a heating duct is rarely worth the effort. But where the ducts emerge from the furnace to begin their journey through your house, a layer of 2-inch fiberglass blanket insulation can cut your heating bill by several additional percentage points and pay back its investment in just about two years.

Before you install insulation around your ducts, run air through them and check for leaks by passing your hands around the joints. Seal any leaks you may find with duct tape.

Ducts are insulated with blankets and batts, and ordinarily have no vapor barrier. However, if your ducts are used in air conditioning as well as in heating, you should make sure to use insulation *with* a vapor barrier, and install it with the barrier facing to the outside, away from the surface of the ducts.

Without crushing the insulation, wrap the blankets securely around the ducts. Where the blankets meet, and at all the wrapping junctures, seal the insulation tightly with duct tape. Make sure you are not blocking any air intakes, and keep both tape and insulation out of contact with light fixtures, chimneys, and flue pipes.

If your ducts are located between joists—as, for example, at the ceiling of the basement—cut strips of batt or blanket insulation long enough to reach across not only the duct, but the adjacent joists as well. Cover the duct securely, and nail or staple the insulation to the outside edge of the joists.

■ Pipe insulation: The same principles apply to pipe heating systems as to forced air systems. You will want to insulate those pipes that are exposed in unheated areas of your house. (See illustration, page 63).

Duct Insulation

1. Run air through the ducts and check for leaks by passing your hands around joists. If you find any leaks, seal them with duct tape.

2. Without crushing the insulation, wrap the blankets around the ducts. Remember to keep the vapor barrier on the outside if the ducts are used for air conditioning instead of or in addition to heating.

3. Where the blankets meet, and at the wrapping junctures, seal the insulation tightly with duct tape. Make sure you aren't blocking any air intakes, and keep the tape and insulation out of contact with light fixtures, chimneys, and flue pipes.

4. Cut the insulation that reaches beyond the duct so that you have two parallel cuts, extending as far as the duct itself is high. Fold the flap of insulation you have just created over the end of the duct and tape it closed securely. Fold the two remaining ends of insulation over the first, and tape them. If you have more insulation on either flap than fits easily over the end of the duct, simply cut it off.

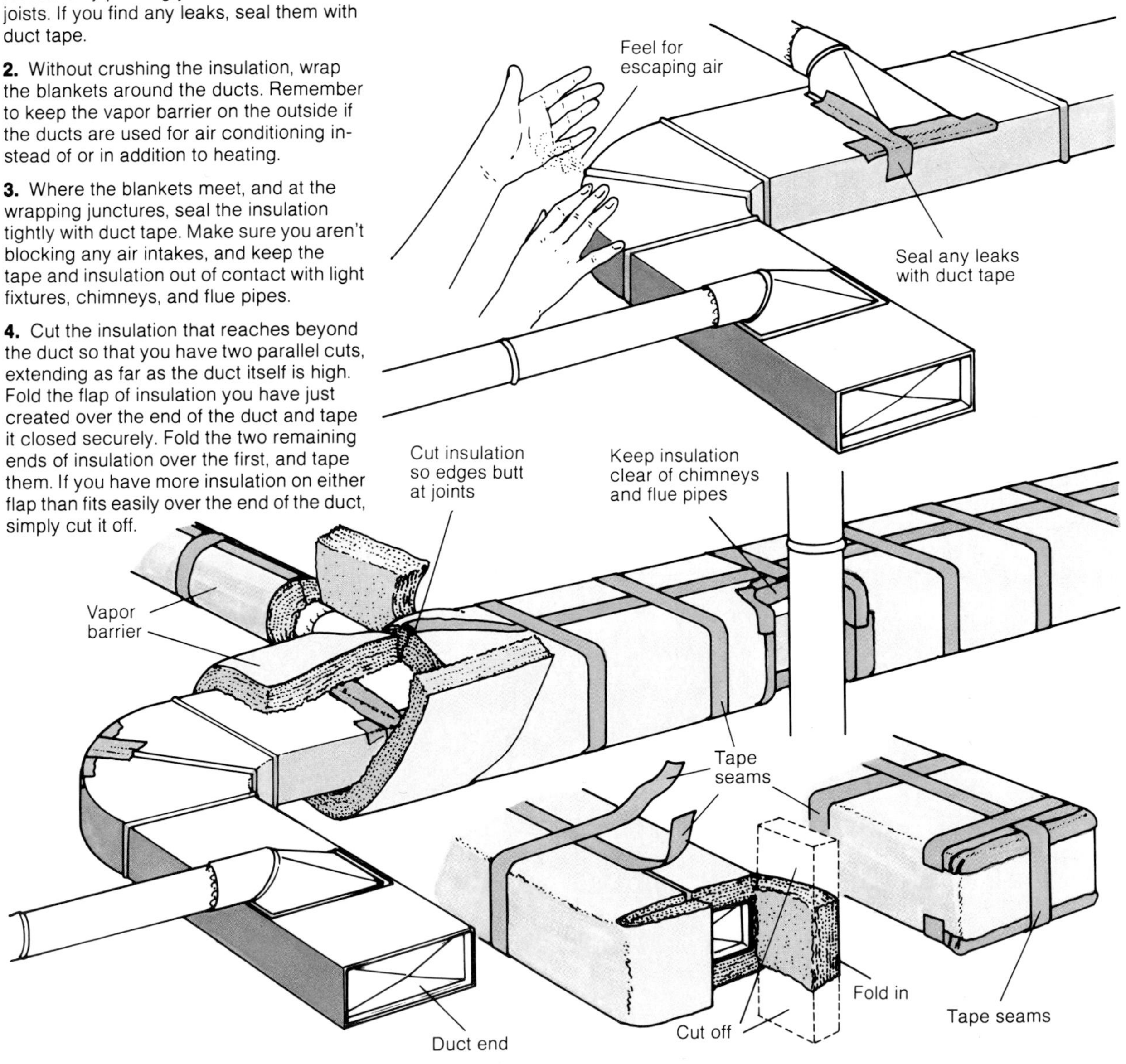

Cooling Systems

Logically, since insulation keeps heat *in* so effectively in cold weather, it ought to keep it *out* just as effectively in hot weather. In general it does—so to get the most significant savings possible in climates that get very hot during part of the year, it's worthwhile insulating to keep the heat out and the cool in.

The whole purpose of insulation is to retard the flow of heat from a warm area to a cool one. So to be effective, insulation must have some heat flow to retard. In winter in most parts of the United States, there is a great deal of heat flow for the insulation to stop—temperatures are considerably lower outdoors than indoors.

In summer, however, there is less heat flow to obstruct; the temperature difference between inside and outside is not as great, and sometimes (for instance, at night) the outside temperature may even be lower than the inside one.

As a result, in most parts of the United States, insulation is more cost-effective when it retains heat in winter than when it retains coolness in summer (although you will receive some benefit from both).

Unless otherwise noted, the following information applies to both insulated and uninsulated houses. However, for purposes of illustration, we are assuming that your house has been insulated to the recommended levels for your particular region.

■ **Circulation:** Conditioned air is comfortable air. In warm weather, conditioning may involve dehumidifying and circulating the air.

Air that circulates evens out humidity, draws off body heat, balances temperatures even in a confined space, and freshens the surrounding air. Each of these processes makes you *feel* cooler, although none may actually reduce the ambient temperature.

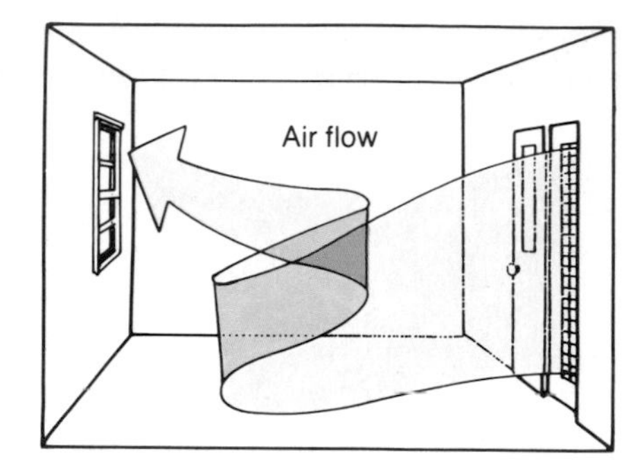

When the weather is hot, often you can condition indoor air more effectively by providing adequate ventilation than by turning on an energy-dependent air conditioner. Instead of automatically reaching for the air-conditioner switch, you might consider that a gentle movement of air through the open doors and windows of the house can go a long way toward providing indoor comfort. And this cooling air is free (as long as the air conditioner isn't running at the same time, of course).

What Ventilation Does

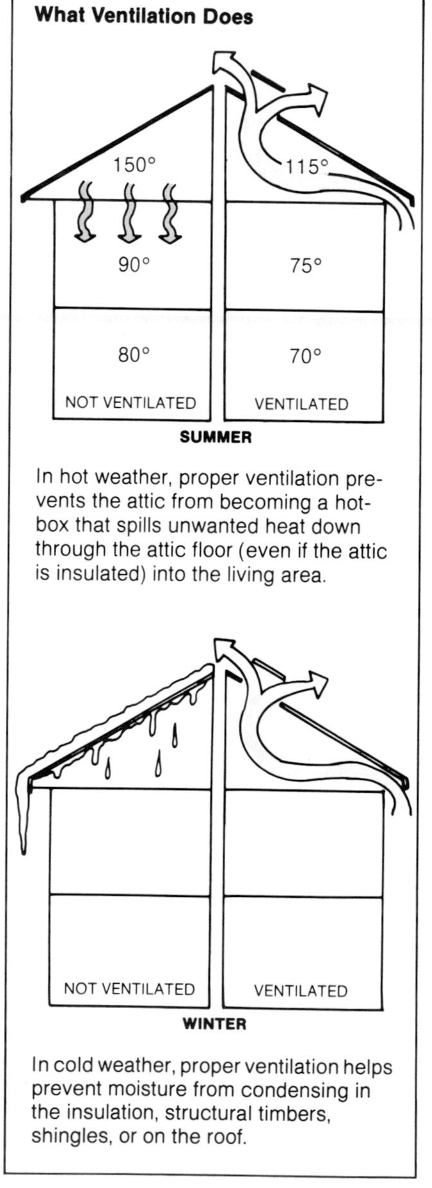

In hot weather, proper ventilation prevents the attic from becoming a hotbox that spills unwanted heat down through the attic floor (even if the attic is insulated) into the living area.

In cold weather, proper ventilation helps prevent moisture from condensing in the insulation, structural timbers, shingles, or on the roof.

Kinds of Ventilators

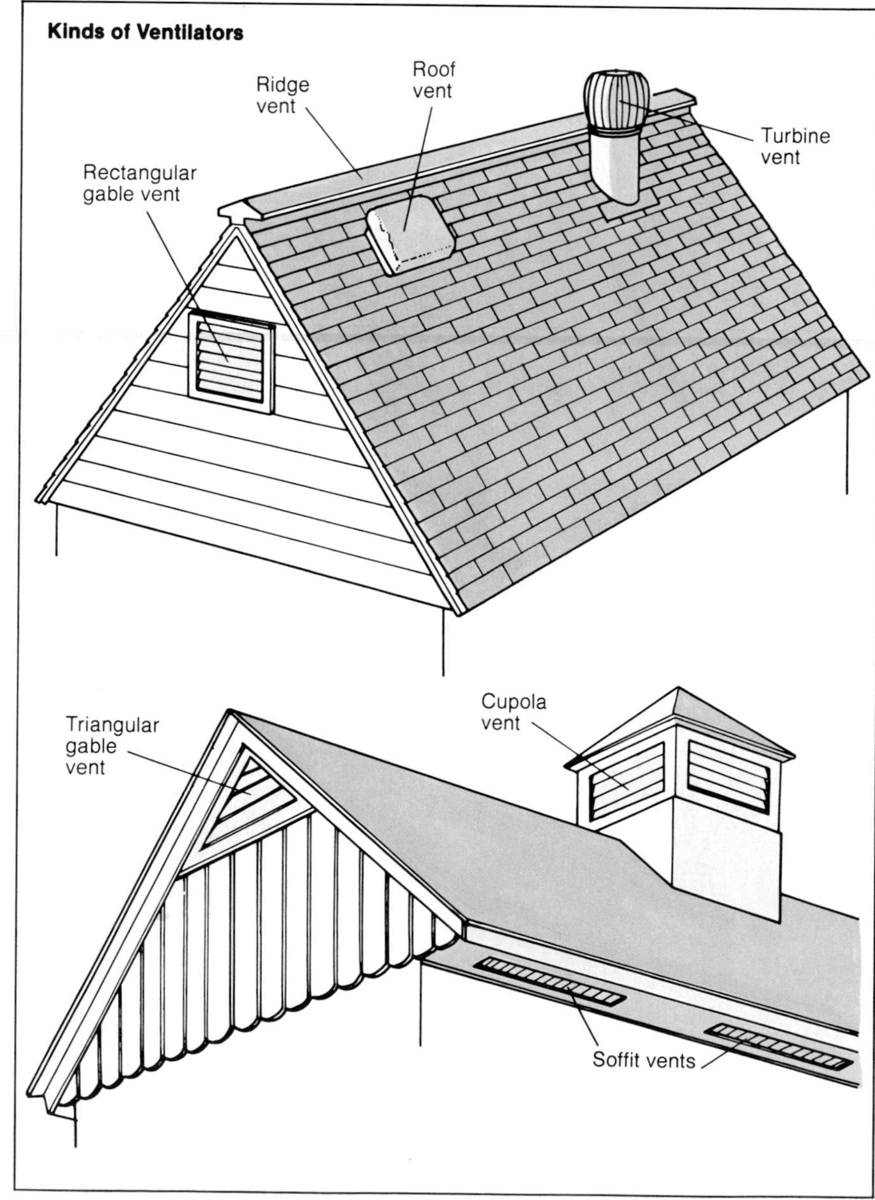

Ventilation

Virtually every house has some venting in or near the roof, although most really don't have as much as they should. The purpose of roof-area vents is to prevent moisture from being trapped at the top of the house, where it can rot the plaster, insulation, and wood; and blister the paint.

In general, a properly insulated attic will have about 1 square foot of ventilation for every 300 square feet of ceiling area. Because the insulation's vapor barrier keeps household moisture out of the attic, the *less* insulation you have, the *more* venting you need, since proper ventilation helps keep moisture from condensing. In any case, it's almost impossible to have too much venting in your attic or crawlspace.

There are many kinds of vents, distinguished mostly by their positions in the attic or crawlspace. The most common venting arrangement consists of a full length of soffit vents and two gable vents (rectangular or triangular), one at either end of the roof's peak. Vents at the roof's ridge are also common.

Most vents simply permit air to flow through themselves and the attic, without providing anything more than their presence to encourage this air flow. The turbine vent is an exception—the lightest breeze makes its blades begin to spin, producing a slight low pressure inside the attic and actually drawing air through the space. Motorized ventilators are also available.

Roof vents are easy to install; you need only measure and cut out a piece of roof or wall that's the same size as the vent, and then insert the vent in the cut-out portion. A system of 2 by 4 inch headers should frame any hole you cut. Then, when you've placed the vent according to the manufacturer's instructions, caulk and flash the area around it to keep the roof from leaking (see "Caulking," pages 28-31).

How Vents Work

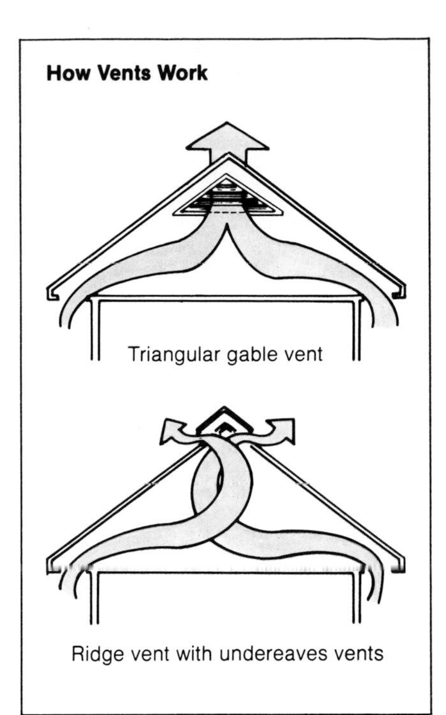
Triangular gable vent

Ridge vent with undereaves vents

Installing a Turbine (or Other Roof Top Ventilation)

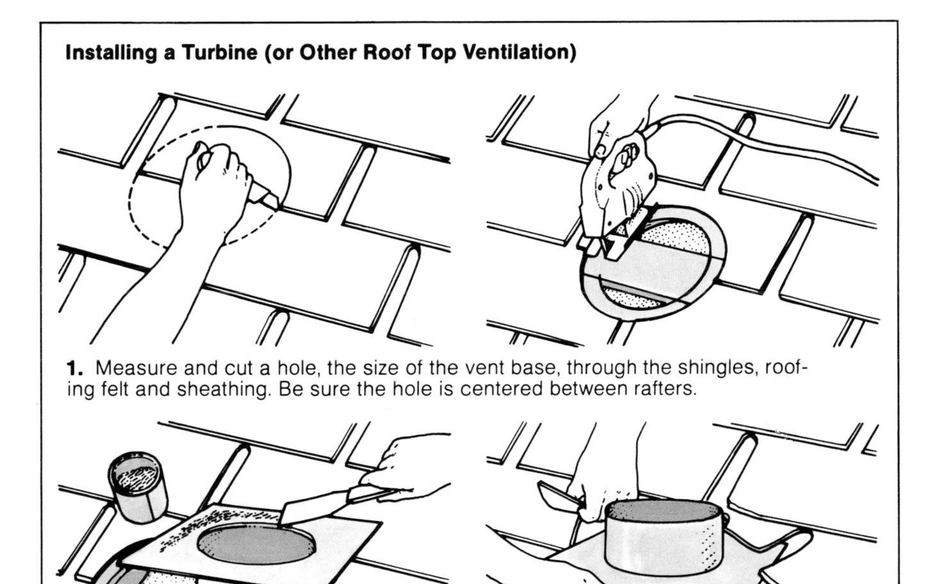
1. Measure and cut a hole, the size of the vent base, through the shingles, roofing felt and sheathing. Be sure the hole is centered between rafters.

2. Apply roofing cement or caulking to the underside of the vent base flange and slide it under the surrounding shingles so the base is centered over the hole.

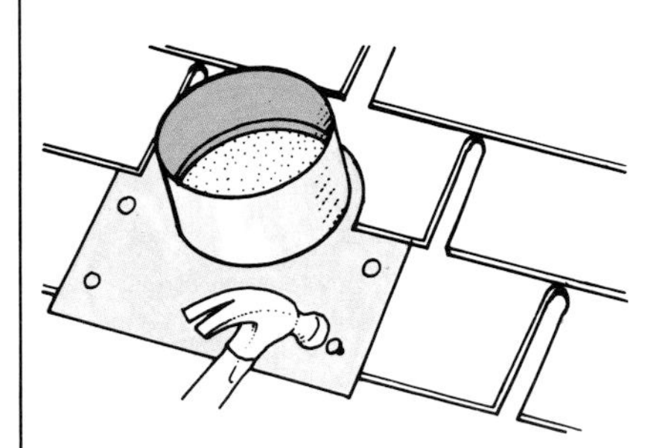
3. Nail the flange in place. Replace shingles if necessary and caulk the edge of the flange and nail heads.

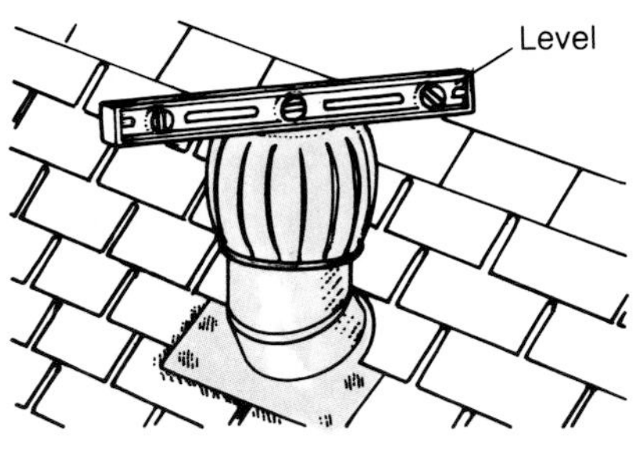

4. Set the turbine (or other vent top) on the base. Level and secure it as shown on its instruction sheet.

Installing a Triangular Gable Vent

1. Cut the hole the size indicated on the manufacturer's insruction sheet. Be sure to allow for the 2 × 4 header.

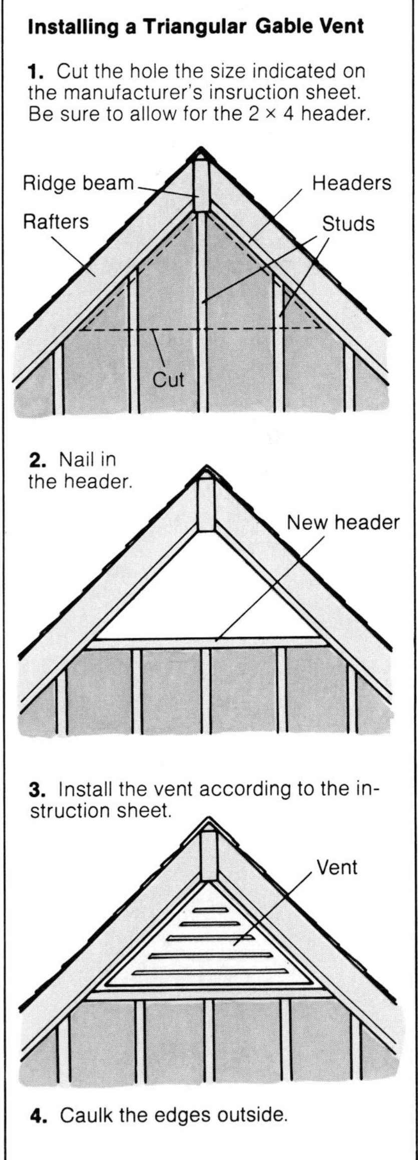

2. Nail in the header.

3. Install the vent according to the instruction sheet.

4. Caulk the edges outside.

Fans

We must admit that simply opening your doors and windows to get circulating air does pose a problem: The air flow these openings provide may be sporadic and gusty—or nonexistent on particularly hot days. This unsystematic method leaves your comfort up to the winds of chance. Fans, on the other hand, provide a more determined cooling effect by propelling, sucking, or forcing air through the living space.

All fans have the same general purpose, and all function in the same general way. So feel free to choose from among large or small; portable or installed; oscillating, tilting, or stable; and window or table fans.

■ **Whole-house attic fan:** The whole-house attic fan is usually an excellent investment. It can increase the comfort level throughout your house, satisfy all your cooling needs on many days, and give your central air conditioner a good, energy-conserving boost.

Whole-house attic fans range in price from about $100 to $400 (not including installation and any of a variety of accessories you might want, such as shutters). If you can get another person to help you, you can install this fan fairly easily.

Whole-house attic fans require that the house and attic be open to each other. Otherwise the living area will not receive the ventilating benefits the fan provides. However, you don't want your fan to stir up loose fill insulation or fiberglass dust from your attic; so if you're thinking about installing a whole-house attic fan, first discuss this point with a conservation analyst from your local utility company.

■ **Installing a whole-house attic fan:** There are three main ways to install a whole-house attic fan, illustrated below. To make use of any of these methods, you must have two air passages—one from the living space to the fan, and another from the fan to the outside. You may be able to use an existing opening, such as a stairway or door, or you may have to cut both openings. If major wiring is required, don't do it yourself—hire an electrician.

When installing the fan, always follow the manufacturer's instructions. The step-by-step instructions that follow are only a general guide.

Once in operation, the fan will draw air from all parts of the house. Be sure that at least some windows or doors are open before you turn on the fan, or your pilot lights may go out, and soot may be drawn from your chimney. If you want the fan to circulate air strongly in one area, close off the remaining rooms.

A note on safety: If children or pets are at all likely to come near this fan, install some sort of barrier around it. The blades can do severe and permanent damage to the person or animal who stumbles into them, and there's no reason to turn your comfort into a disaster.

How the Whole-house Attic Fan Circulates Air

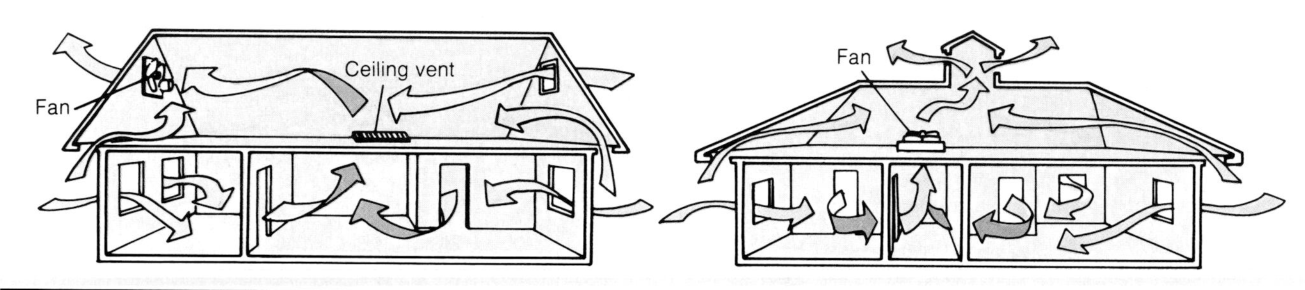

The Whole-house Attic Fan and Three Installation Positions

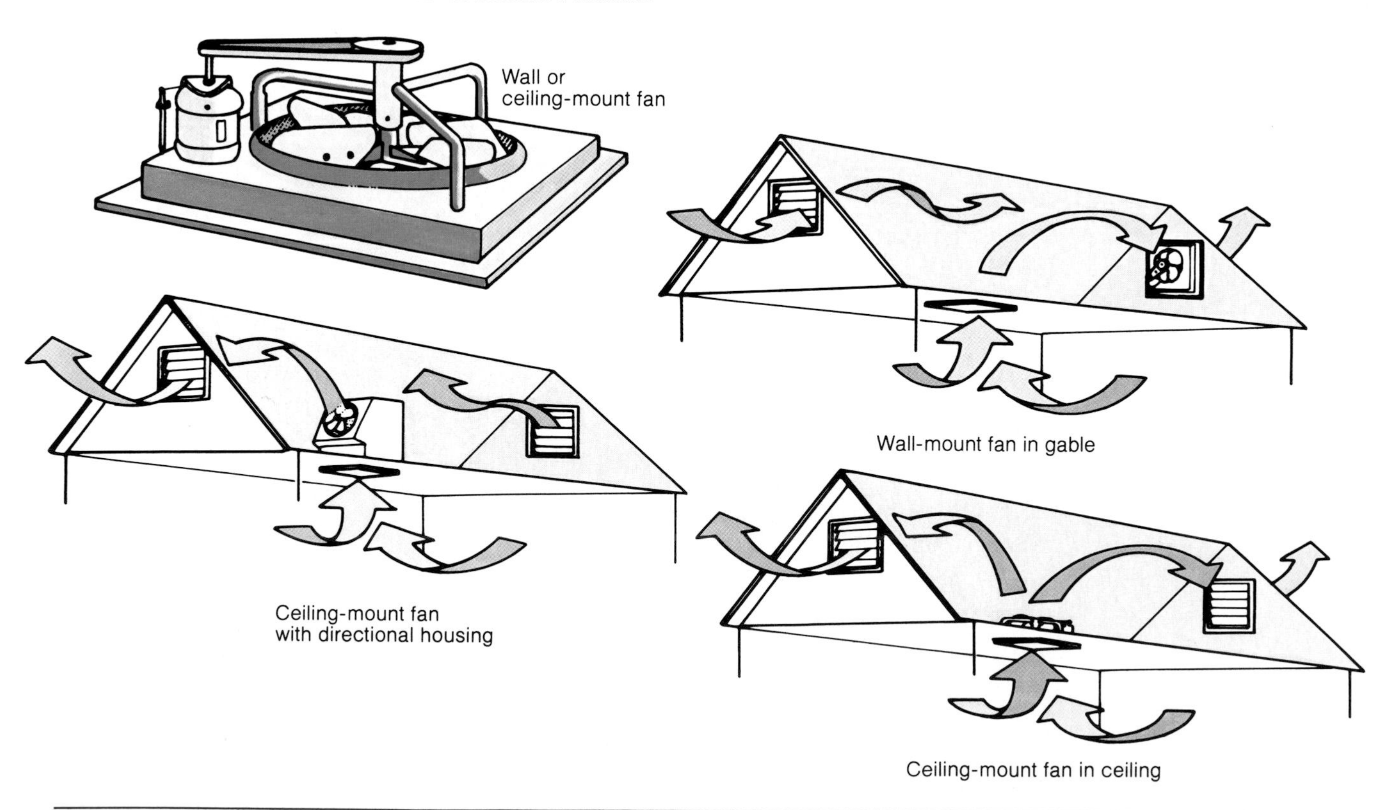

How to Install a Whole-house Attic Fan

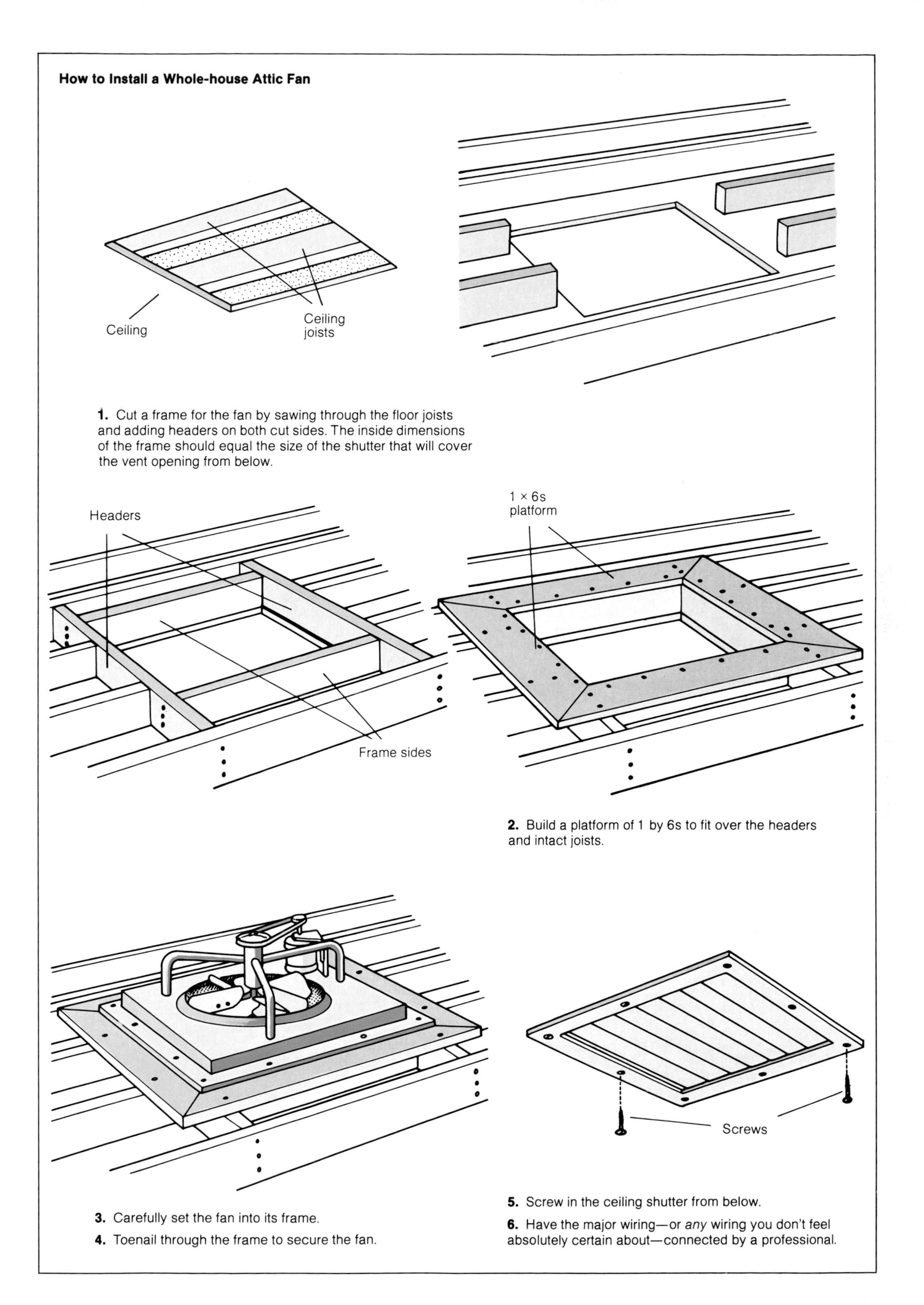

1. Cut a frame for the fan by sawing through the floor joists and adding headers on both cut sides. The inside dimensions of the frame should equal the size of the shutter that will cover the vent opening from below.

2. Build a platform of 1 by 6s to fit over the headers and intact joists.

3. Carefully set the fan into its frame.

4. Toenail through the frame to secure the fan.

5. Screw in the ceiling shutter from below.

6. Have the major wiring—or *any* wiring you don't feel absolutely certain about—connected by a professional.

■ **Attic fans:** Don't confuse a *whole-house* attic fan with a *plain* attic fan—their purposes are quite different. During hot weather, an attic can become extremely warm, especially if it is well-insulated and holds in heat. The built-up heat may leak down into your living space, reducing your comfort; if you run your air conditioner to combat the extra warmth, you end up increasing your utility bill.

The main value of attic fans is in releasing pent-up heat, and this end can usually be better accomplished with adequate venting (see page 58). But you can use an attic fan to create additional circulation in your attic (the attic is still closed off from the living area). The fan propels the heated air out through a system of attic vents.

Installing an Attic Fan

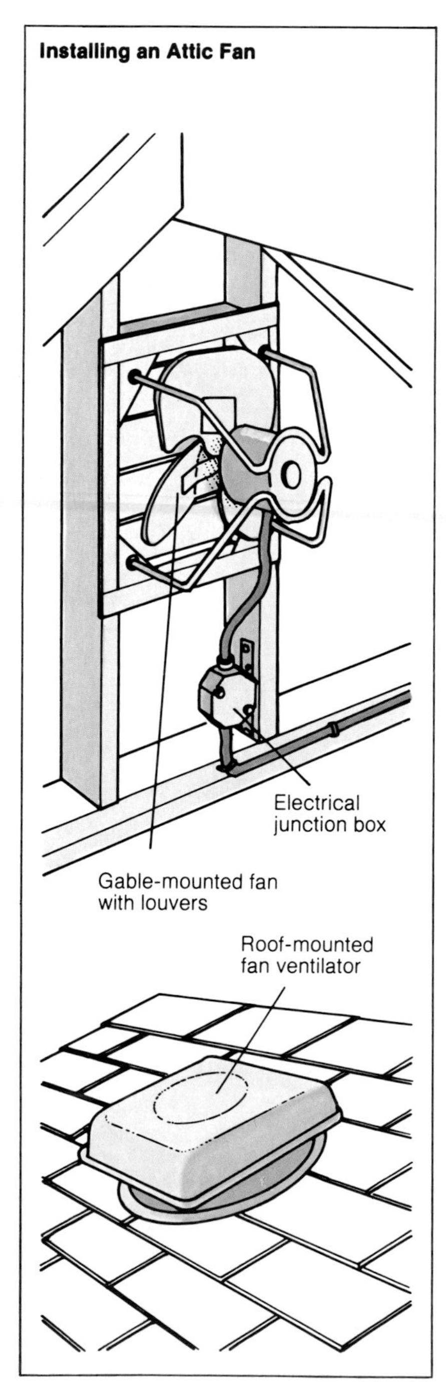

Air Conditioners

Central air conditioning works by passing air through a chilling system roughly similar to that of a refrigerator. It works simply and directly, but expensively. **Room** air conditioning is not as expensive, but it still adds significantly to your utility bill. Despite the rising costs of the fuel needed to operate them, these methods remain popular because of their simplicity, directness, and effectiveness. If you already own either one of these cooling systems, take good care of it and it will give you better service for a longer period of time at lower cost.

■ **Air conditioner maintenance:** To find out how to maintain your air conditioner on your own, talk to your dealer or service person. In general, you should clean a central air conditioner's condenser coils annually, and clean or replace its air filters every month or two, depending on the frequency and intensity with which you use the unit. You can clean the coils with an ordinary vacuum cleaner, sucking up the dust that has accumulated. If the hose on your cleaner will not fit between the coils, you can also reverse the vacuum and blow the dirt away. And you can reduce the air conditioner's fuel waste by providing some shade or cover for the condenser outside (of course, make sure that the shading doesn't block the passage of air into or around the condenser).

Like your furnace, your air conditioner should receive an annual checkup from a trained service person. The best time of year for the checkup is spring, just before the air conditioner will be in frequent use. Have the service person oil all unsealed bearings; tighten belts; check for and repair refrigerant fluid leaks, and add refrigerant if necessary; flush the evaporator drain line; and secure all electrical connections. In addition, if your heating and cooling systems use the same ducts, the dampers should be adjusted each time you change from one system to the other, since each one requires a different setting.

■ **Use your air conditioner wisely:** Learning to use your air conditioner effectively can help reduce your energy bills a great deal. For starters, keep your windows closed while the air conditioner is on; close your drapes to keep the sun out of any room you want to cool; and don't run your air conditioner when you're going to be gone for any length of time. But to make your air conditioner truly effective, remember that its principal purpose is to *condition* the air, not just to cool it. Use it only when you must.

Water Heaters

Heating water consumes the second largest amount of energy in your house. About 20 percent of the energy used by the typical American house goes to heat water, compared with about 70 percent for space heating, and about 10 percent for lighting, cooking, running appliances, and so forth.

Frustratingly, there are only a few measures you can take to reduce the waste; however, all of these are easy, inexpensive, and will pay back your investment within a few years—usually one or two.

The easiest and least expensive conservation measure involves simply lowering the thermostat setting on your hot-water tank. (See page 15). When the temperature of the water is reduced, the difference between the water tank temperature and that of the surrounding air is also reduced. This means that less heat from the tank will be lost to the cooler air surrounding it. There are other less significant savings. For instance, dishwashers and washing machines that are regulated by a lower thermostatic setting will continue to use the water-temperature mix for which they have been set, only now they will use less hot water. Occasionally, dishes may appear spotted after the cooler wash, but the Department of Energy still recommends keeping dishwashing temperatures as low as 120°.

Some dishwashers come equipped with both built-in heat sensors and with heaters that raise the temperature of the water to 140° during the final rinse cycle at the rate of about 1 degree every 2 minutes, no matter what the temperature of the water is as it enters the dishwasher. If your dishwasher has such a sensor, it makes little sense to keep

Cleaning Air Conditioners Condenser Coils

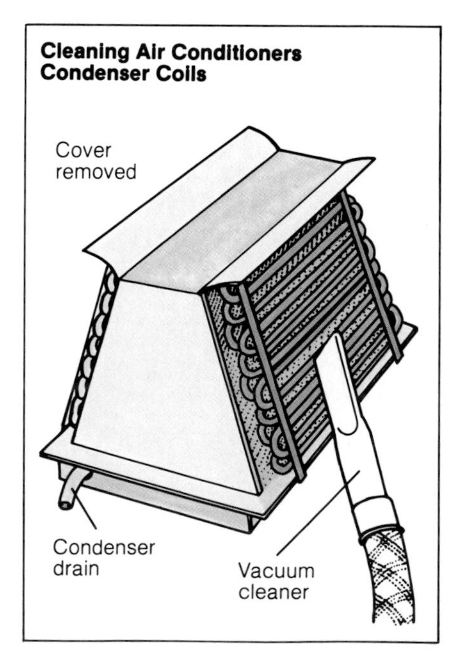

it running for the extra 40 minutes or more that an initial 120° water temperature will elicit, in which case you may want to keep your water heater temperature high. If you don't know whether your machine is equipped with a heat sensor, ask your service person or call your dealer.

Water heater thermostats typically are set at about 140°, with some as high as 150° to 160°. But almost no household use requires water temperatures above 120° (except for those dishwashers mentioned above), and 110° is satisfactory for most hot-water uses. Whether you set your heater's thermostat at a high figure to accommodate your dishwasher or set it lower and wash your dishes by hand is up to you. You should know, however, that the cost of heating 100 gallons of water in your heater to 160° from an inflow temperature of 50° is nearly 30 percent greater than heating those same 100 gallons to 120°. The dollar translation for you will depend on which fuel you use to heat your water, and what price you pay for it. But you can pay as much as $5 per month for every 10 degrees your water is heated above 120°.

■ **Insulating your water heater:** The next step is to add insulation to your hot-water tank. If it feels warm when you touch its sides, especially if your tank is in an unheated basement or on an open back porch, this is a must-do project.

All commercially available hot-water tanks are made with some interior insulation—electric heaters have 2 inches of insulation; oil- and natural gas-fired heaters have about 1 inch. These heaters tend to lose about 20 to 35 percent of their heat through their walls. A single, 3½-inch-thick roll of R-11 fiberglass blanket, sealed with duct tape and costing about $20, will be sufficient to wrap a 50-gallon tank, and will pay for itself in one or two years. You can also buy precut kits, but the job is so simple and cheap that you can cut the insulation yourself with little work.

When insulating your water tank: (1) do not block the pilot light or access to controls on any gas heater; (2) do not cover the top of a gas heater or put the insulation anywhere near the burner; (3) if you don't expect to use hot water for a protracted period of time, make sure your gas or oil heater has a pressure-relief valve. Otherwise, the additional insulation may make the water overheat. Many states have codes that require such valves—if you aren't sure about your state, call your public utility company; and (4) make a note of your serial number in case you need to get a part for your particular model.

When insulating your hot-water heater, you might also insulate the first ten feet or so of hot- and cold-water lines coming from the tank. This will minimize how long you'll have to wait for the water that comes out of your faucets to reach the desired temperature. Although such insulation may not save you much in terms of either money or heat loss, if your pipes pass through an unheated area where they might freeze in winter, this task may be worth your while. To save money, you can use a lesser insulation, such as a combination vinyl foam/aluminum foil self-adhesive tape, on the pipes, and a superior-quality insulation on the tank. There is also an insulating tape available for use on joists.

■ **Maintenance:** Your water heater is a piece of machinery in a system of its own. Like a furnace or an air conditioner, it should receive a periodic checkup from a trained service person who will de-lime the tank in hard-water areas, clean the burner, or clean the electrodes, depending on whether your heater runs on gas, oil, or electricity. And every three or four months you should drain a gallon or two of water from the bottom of the tank. This will remove the sediment that, if left in the tank, will keep the water insulated from the burner flame.

How to Insulate a Water Tank

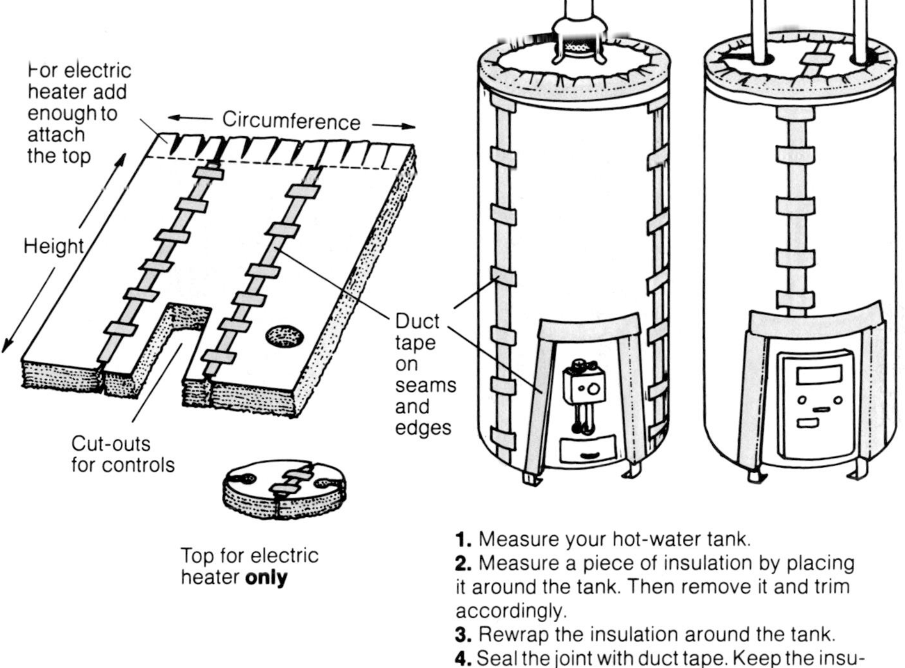

1. Measure your hot-water tank.
2. Measure a piece of insulation by placing it around the tank. Then remove it and trim accordingly.
3. Rewrap the insulation around the tank.
4. Seal the joint with duct tape. Keep the insulation away from the pilot light, if there is one.

Pipe Insulation

Snap pre-slit foam sleeves onto pipes and twist to secure . . .

or wrap pipe with adhesive-backed foam tape.

Drain sediment from water heater

Light dimmer

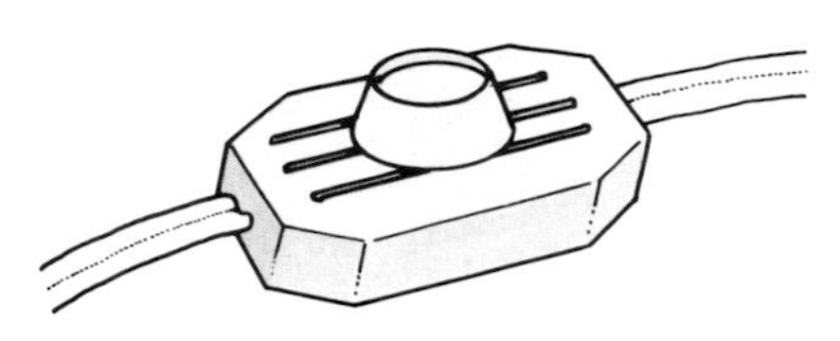
Cord dimmer switch

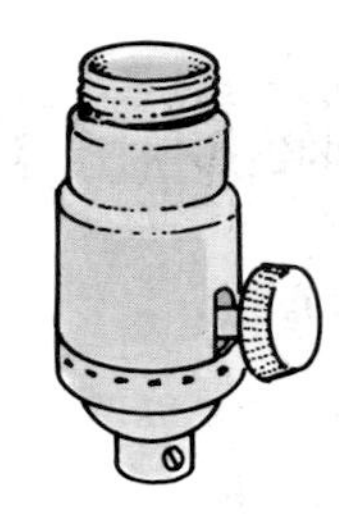
Dimmer socket

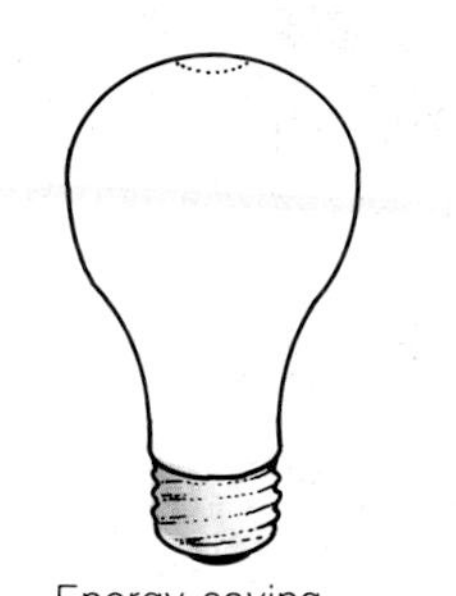
Energy-saving
incandescent light bulb

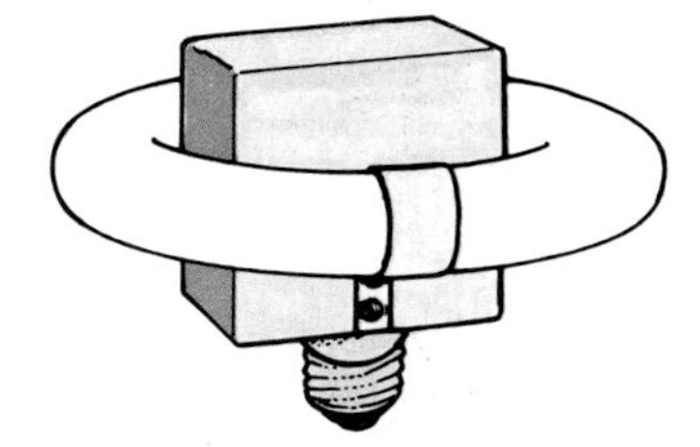
Lamp and fixture converter

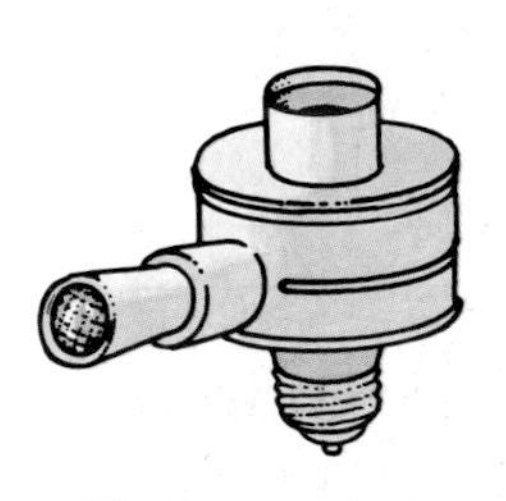
Photoelectric light cell

Energy-Saving Gadgets

■ **Light dimmers:** Light dimmers replace regular light switches and allow you to control the wattage without replacing light bulbs. By turning the dial, you can direct the light level from bright to dim, depending on the level of light you need at the time. One dimmer probably will not make much difference in your electricity bill, but if you install dimmers throughout your house and use the lower wattage frequently, you may realize some savings. Some people purchase dimmers simply because they like the effect—low lighting for dining, for instance. A single dimmer costs about $7 to $12. Installation instructions come with it. Follow these instructions meticulously when dealing with the wiring, and be careful.

■ **Cord dimmer switches:** Cord dimmer switches attach to lamp cords and offer two levels of light—bright and dim. As with wall dimmer switches, you probably won't realize great energy or monetary savings unless you use them throughout the house. Depending on how often you use it, a lower setting will extend the life of an incandescent light bulb. Cost is about $7 to $8 each. Installation is simple.

■ **Dimmer sockets:** Lamp dimmer sockets are another way to control the wattage and resulting amount of light. They can be inserted in any standard lamp, and they offer the full range of control instead of just two settings. Cost is about $7 to $8 each. However, whether you use wall switches, lamp-cord attachments, or sockets, dimmers are practical only when you want lighting that ranges from bright to less bright. And you would get almost the same flexibility and savings with three-way light bulbs.

■ **Energy-saving incandescent light bulbs:** These are just like ordinary incandescent light bulbs, except for an inside coating that reduces the amount of electrical energy used but doesn't affect the amount of light produced. Replacing the 100-watt conventional bulbs in directional lighting fixtures with 50-watt R-20 reflector bulbs can save you as much as 50 percent of your lighting costs. And you can change the 100-watt or 150-watt reflector bulbs in ceiling fixtures to 75-watt ER bulbs and still have adequate light while using less energy. A 4-watt clear night-light bulb (about $1.50) will give the same amount of light as a regular, frosted 7-watt bulb.

■ **Lamp and fixture converters:** When you screw converters or adapters into standard lamp or fixture sockets, the sockets are ready to take fluorescent lighting. New circline fluorescent tubes are available in standard sizes for these converters. The circline fluorescent tube in a lamp uses 60 percent less energy than the ordinary incandescent light bulb, since it requires only 29 watts to burn as brightly as a 75-watt incandescent bulb. It also has a rated life of 12,000 hours, compared with 750 hours for an incandescent bulb. The cost of a converter or adapter ranges from approximately $17 to $25. Some ceiling light converters are designed to be complete decorator fixtures that will fit existing mounting boxes. These cost $30 to $35.

■ **Photoelectric light cells:** If you keep an outside security light on through the night, perhaps you find that you forget to turn it off first thing in the morning. A small investment of $7.50 to $12 will get you a photoelectric cell that *won't* forget—it will automatically turn the light on at dusk and off again at dawn, a 50 percent energy and cost saving. Just screw this device into any conventional light bulb socket.

■ **Outlet and switch gaskets:** If you insert these precut foam gaskets behind the decorative switch plates and electrical outlet plates on your house's outside walls, they will help cut down heat loss from drafts and save you about 3 percent in home heating costs. Priced from $3.50 to $7.50, they are usually sold in packages, with varying numbers of gaskets for each use. With a screwdriver, remove the plates, fit in the gaskets, and replace the plates.

■ **Automatic furnace ignition devices:** An automatic ignition device on your forced-air furnace will light the pilot only when the thermostat calls for heat. It

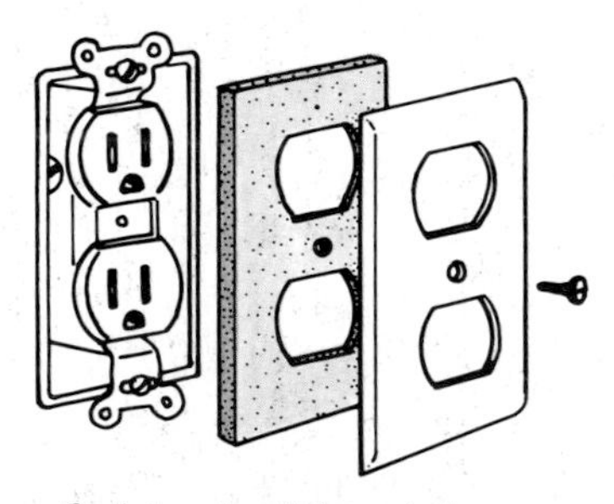
Outlet and switch gasket

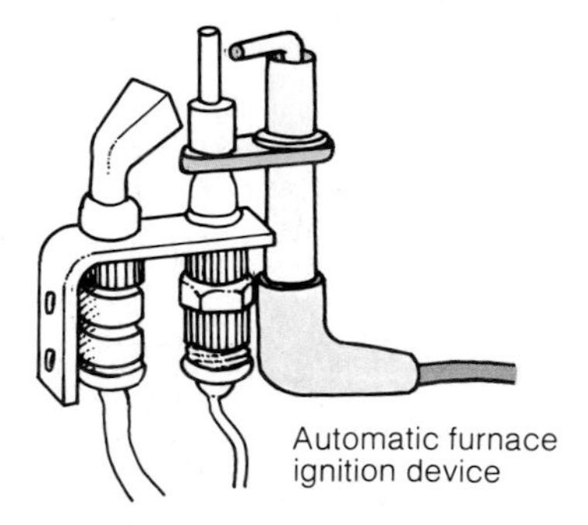
Automatic furnace ignition device

eliminates burning the pilot light 24 hours a day, which saves about 81 therms, or a minimum of $20, a year. Also called an *intermittent pilot system,* it is as safe or safer than your present pilot light. However, it cannot be installed on a wall heater or a gravity-type furnace (the latter has no forced air-the heat rises because hot air rises). Installation costs about $175 and must be done by a licensed contractor.

■ **Clock thermostats:** A clock thermostat on your heating system can save up to 30 percent in costs a year: it automatically turns the heat down when you go to bed, and then turns it up again in the morning when you wake up. You can also use a thermostat for your cooling system—it is equally efficient for that purpose. The combined savings can add up to $75 or more annually. You can install clock thermostats yourself or have them installed professionally. A clock thermostat, or equivalent modifiers and timers, costs about $65, plus installation (this charge can vary considerably from area to area).

■ **Portable heaters:** Portable, plug-in electric heaters *can* save energy and heating costs—but only if you use them instead of your furnace in rooms that get only occasional use. The price depends on the style, size, and number of heat settings; usually it ranges from $20 to $40. Many types can be used either as a heater or, with the fan, as an air circulator.

■ **Clothes dryer heat-recovery devices:** There are at least two heat-recovery systems for clothes dryers. Attaching to the dryer vent, both are designed to direct dryer heat and humidity back into the room during cold weather, and outside during summer months. With most dryers, a heat-recovery system will save 20 therms or 240 kilowatt hours per year —a $5 annual saving for a gas-heated home, or a $12 yearly saving for an electrically heated home. Moisture and contaminants may be introduced inside the house, although a model with a filter will eliminate this danger. Compared with the energy saved, the cost of this device is high.

■ **Energy-saving switch for air conditioners:** An energy-saving switch on a

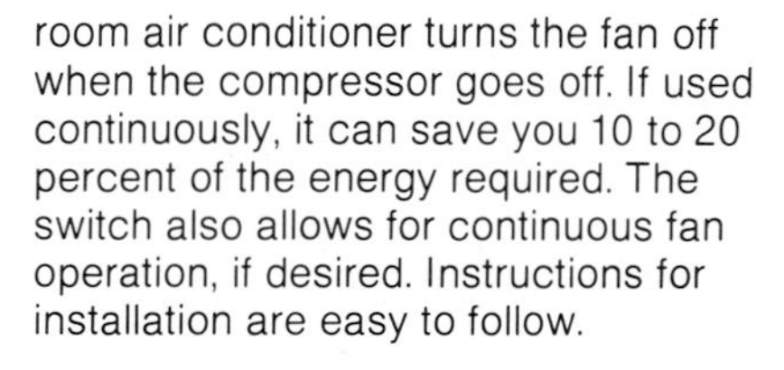

room air conditioner turns the fan off when the compressor goes off. If used continuously, it can save you 10 to 20 percent of the energy required. The switch also allows for continuous fan operation, if desired. Instructions for installation are easy to follow.

■ **Air deflectors:** Molded plastic air deflectors simply direct heat—or cool air —where you want it from your central forced-air heating or cooling system. They allow more efficient use of energy (for example, by keeping drapes from blocking the heat flow), and they make the room generally more comfortable; however, energy and cost savings are minimal. The air deflectors adjust in length from 9 to 14 inches, and are held in place by strong magnets. Cost is modest: $3.50 to $4.50.

■ **Refrigerator-coil brush:** A refrigerator-coil brush does just what its name implies: tapered bristles clean hard-to-get-at coils where your vacuum cleaner or other brushes just can't do the job. A clean refrigerator compressor operates more efficiently than a dirty one, saving up to 10 percent of the electricity used. Cost is approximately $5.

■ **Clogged filter indicators:** It's easy to forget to check your forced-air furnace or air conditioner filter, and clogged filters can cost you up to 5 percent more than need be. Whether your system uses disposable or reusable filters, filter alerts or indicators show you when to change or clean them. They also work with most electric air cleaners. These plastic devices are easy to mount on blower compartments. The cost: $13 to $15.

■ **Shower-head flow restrictors:** Shower-head flow restrictors reduce the flow of hot water from approximately $3\frac{1}{2}$ gallons per minute to $2\frac{1}{2}$ gallons per minute. Since hot-water heating accounts for 20 percent of residential energy consumption, cutting down the flow of hot water in daily showers can result in substantial savings. A wide range of water-saving shower heads and water restrictors is available, so flow rates and energy saved will vary accordingly. Cost runs between $9 and $12.

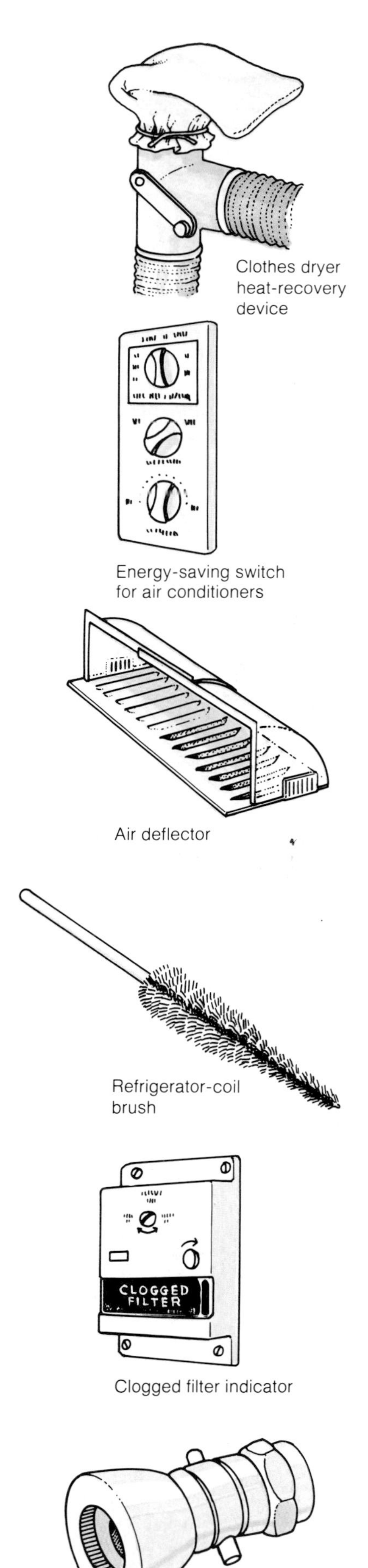

Clothes dryer heat-recovery device

Energy-saving switch for air conditioners

Air deflector

Refrigerator-coil brush

Clogged filter indicator

Shower-head flow restrictor

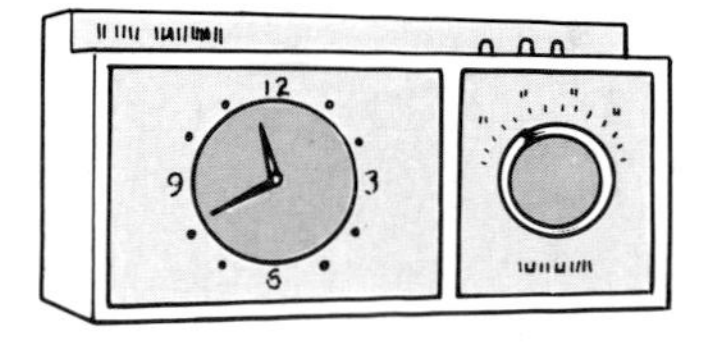

Clock thermostat

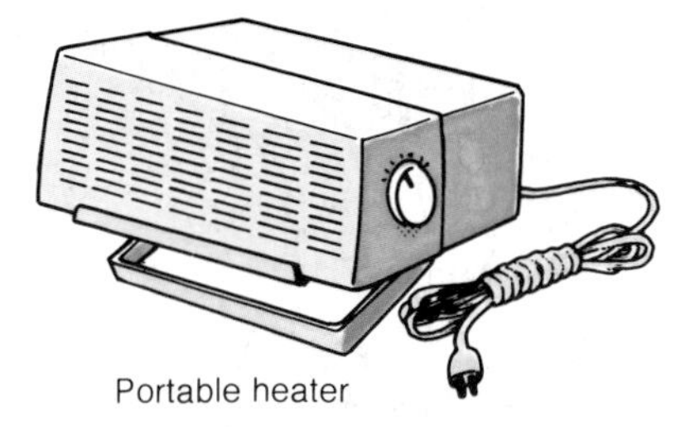

Portable heater

Swimming Pools

Did you ever put on your swimsuit on a blazing hot day and jump into your swimming pool, only to spend the next five minutes gasping from the shock of the freezing water? It isn't really surprising that the water in swimming pools cools down so much: The entire surface of that body of water is exposed to the air, at least when the pool is in use, and it loses heat through convection, radiation, and particularly through evaporation. For every pound of water that evaporates, nearly half a ton of water is cooled by 1°F., requiring more heat to keep the pool water at the desired temperature.

What you want is to be able to enjoy your pool at a comfortable temperature, but without losing all the heat you're paying for. A 1980 Pacific Gas & Electric study suggests five ways to conserve energy in a heated pool:

1. Lower the pool thermostat setting.

2. Narrow the pool-heating season to the warmest months in your area.

3. Turn off the heater when the pool is not in use.

4. If possible, keep wind and shading near the pool to a minimum.

5. Use a pool cover.

Heat Loss from an Uncovered Pool

10%
Losses to ground and other

30%
Radiation to sky

60%
Evaporation, convection, conduction at water surface

An uncovered swimming pool loses heat in three ways—by conduction to the earth around it, by radiation to the air, and by evaporation, which cools the heated water in the pool. A simple swimming pool cover reduces most of the evaporation, which is the primary source of heat loss. It also traps the heat, which slows the radiation of heat to the sky.

A plastic bubble pack pool cover can be rolled up and stored out of your way while your pool is in use (top). When you're finished swimming for the day, or if you plan to be away from home for a while, replace the cover over the entire surface of your pool. You can accomplish this task with or without a hand-winch mechanism (middle). The cover will retain heat and severely reduce evaporation, yielding large savings.

The expense of heating a swimming pool with fuels during high-sun season (May-September) is approximately equal to driving a 20 mpg car 15 thousand miles. A simple plastic vinyl pool cover can reduce this expense to almost nothing. An electric mechanism such as the one shown here makes the task of covering and uncovering the pool very simple.

■ **Pool covers:** Using a pool cover, or pool "blanket," at night and on overcast days when the pool is not in use offers three advantages: (1) The pool cover retains the heat already stored in the water; (2) It retains the water itself, thereby preventing the massive evaporative heat loss; and (3) It can even raise the temperature of the water at the pool's surface—the first 12 inches or so, and more if the filter is in operation—by several degrees. The PG&E study found that the cost of a pool cover often can be recouped in a single year's savings.

Several kinds of pool covers are available, ranging in price from about 25¢ per square foot for simple plastic film, to about $1 per square foot for woven plastic blankets with ultraviolet inhibitors. If you are prepared to pay up to $4 per square foot for your whole installation, you can have an electric winch that rolls and unrolls your cover across the top of your pool for you, and stores it conveniently out of the way while you swim.

Whether or not they are equipped with such devices, all pool covers perform essentially the same function: They cover the pool when it isn't in use. The most common covers are dark or transparent plastic film, clear bubble-pack plastic, and opaque insulated film.

If you expect to use the cover often during daylight hours, choose the clear bubble-pack, since it will allow the pool water to absorb solar heat, while simultaneously insulating it against the loss of that heat. If you expect your pool to be uncovered most of the time on most days, but covered at night, an opaque foam cover will provide the greatest protection against heat loss by convection and radiation. Plain opaque or transparent plastic film is less effective and less expensive, but far better than no protection at all; it will retard evaporation, particularly if it can be secured tightly around the pool's edges.

Unlike the insulated covers, which should last several years, a plain sheet of film may last only a single swimming season and then have to be replaced. In some states, the additional expense of an insulated cover can be more or less balanced out by a tax credit that requires a three-year manufacturer's warranty. Plain covers don't have such a credit.

Once your pool cover is installed, treat it with the respect its manufacturer recommends. In particular, pay attention to the way you store the cover when it is not in use. It should be kept out of the sun at such times, or covered with some other insulating material.

When the cover is lying on top of your pool the water underneath absorbs and diffuses the sun's heat and keeps the cover itself cool. But when the cover is off your pool, if it lies around in the sun for very long it is liable to crack and blister, which will severely reduce its effectiveness as well as the length of time it will be useful.

■ **Pool heaters:** To create the heat that pool covers will conserve, you can use a conventional gas or electric pool heater. As with any heating unit, it is important to maintain it in good condition. All the guidelines for servicing your home heating unit apply to your pool heater. The efforts you make to be conscious of and conserve the energy you use indoors and out can save you money.

If, after you have taken all possible conservation steps, you want to explore further energy-saving options, you may want to consider a solar pool heating system to augment the water's natural tendency to collect solar heat. Solar pool heating systems are discussed on pages 90-91.

Solar energy is simply the sun's power. Collected, stored, and distributed in a controlled fashion, it can help reduce your consumption of gas, oil, and electricity. You can use the sun's energy in many ways, and some cost very little money.

Solar Energy

For all the enormous quantities of literature and hardware that are suddenly available on the subject, you'd think solar energy was the most complicated new toy since nuclear physics. Actually, it's not new; and it can be as simple as sunlight passing through a window.

Solar **energy** is simply the heat radiated by the sun. The **technology** of solar energy is concerned with its **application**—how to tap and use this unlimited, renewable, nonpolluting, and very available source of energy. Solar energy **systems** are the mechanisms and techniques that have been developed to harness and use this energy.

Although someday we may be able to convert the sun's energy directly into electricity at competitive costs (by means of an emerging science called **photovoltaics**), for the moment the domestic use of solar technology is largely confined to two areas: heating the living **space** within a house; and heating **water**—for domestic use such as bathing, cooking, and drinking, and for swimming pools or hot tubs.

While solar energy has been with us for a long time, the technology that allows us to control its use intentionally and productively has a shorter history (see next page). In some countries, the use of basic solar energy principles has been an integral part of the culture and of architectural design for thousands of years; but this has not been the case in our country. While there have been spurts of growth and development in solar energy technology, most home heating systems still depend primarily on fossil fuels.

Because the design of the houses we live in today was based on the assumption that cheap fuel would be readily available, most of this book has been devoted to increasing the energy efficiency of our existing structures and systems. But with the rapidly increasing cost of fossil fuels, solar technology has received renewed interest; research and installations are being subsidized by federal and state governments. One result of all the attention solar energy has received of late is the wealth of emphatic opinion about its value: Whether it is effective, whether it is *cost*-effective; whether it should or should not be developed further, and if so for what sorts of applications; whether the attention devoted to solar energy development and the research funds spent on it would be better used in other areas.

But while the controversies rage, some facts do stand out. A solar heating system does provide independence from the conventional energy grid, but even with the subsidies and rising fuel costs, the *equipment and installation* of a solar water or space heating system is still more expensive than the equipment and installation of a conventional system. The annual savings in solar operational costs must be measured against the number of years it will take to pay off those initial costs. And until fuel costs rise even higher and the expense of solar equipment and installations is reduced, retrofitting a house with a commercial solar system will not be an easy answer for everyone, and not necessarily an obvious cost-effective step.

It is also true, however, that solar technology *does* exist, and in the pages that follow we will explain what it is and how the various systems work. From these descriptions and illustrations you will be able to determine which aspects of this field may be applicable to your particular situation.

For all the differences among solar energy systems, they all share three primary functions:

1. To collect the sun's heat;
2. To store the collected heat; and
3. To distribute the stored heat in a controlled fashion.

The History of Solar Energy Systems

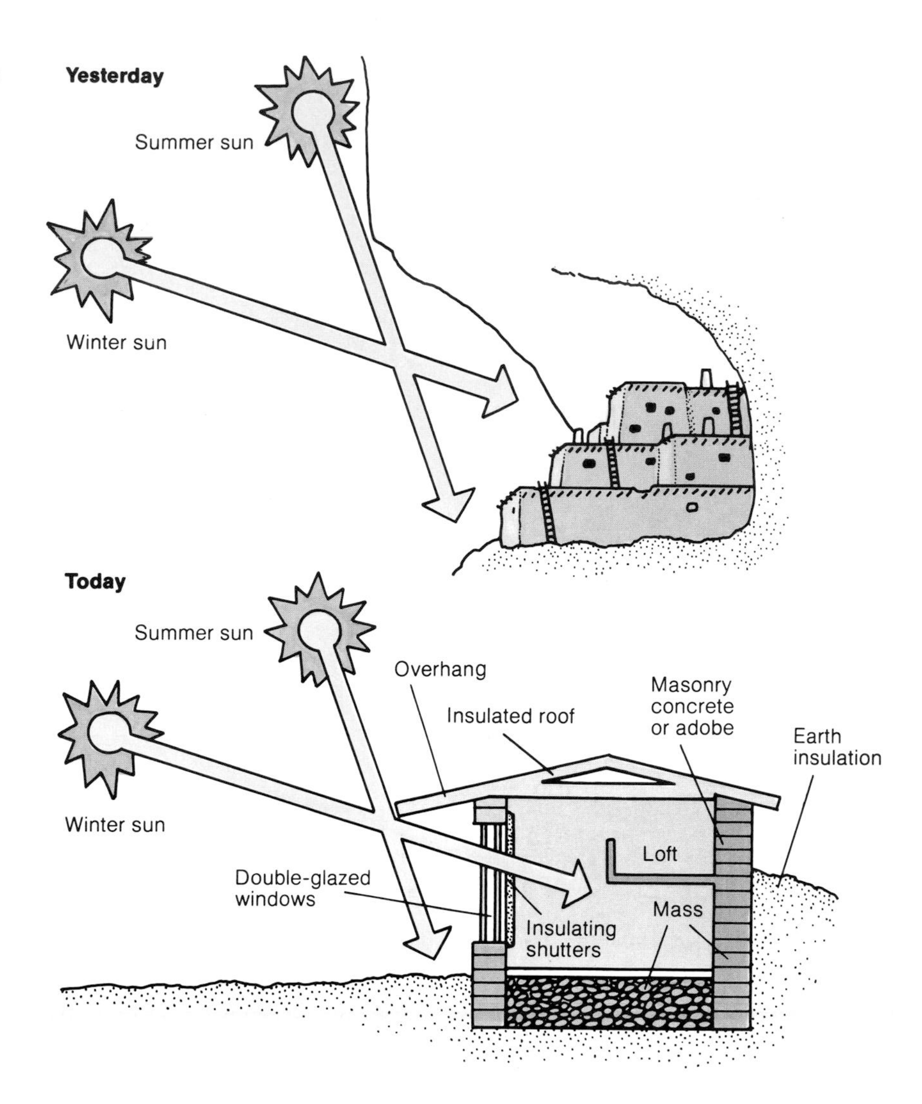

The *presence* of solar energy has never been in doubt. The Pueblo Indians of North America used their adobe huts as mass to absorb that energy during the heat of the day, and to radiate it back to the inhabitants during the chill desert nights. In about 400 B.C., the Greek philosopher Socrates recommended that houses be built with their south sides raised to embrace the sun's heat, and their north sides lowered to keep cold winter winds at bay.

The scientific application of solar energy was under way by 1774, when Joseph Priestley discovered oxygen by focusing the sun's rays onto mercuric oxide; and Antoine Lavoisier, known as the father of modern chemistry, also used large glass lenses to focus the sun's power in the 1770s.

But as Ken Butti and John Perlin point out in their book, *A Golden Thread*, the first significant step in making solar power commercially available did not take place until April 28, 1891, when Clarence M. Kemp, a manufacturer from Baltimore, Maryland, patented his Climax solar water heater.

The Climax, designed to be mounted on a roof or wall, was functionally identical to solar water heaters made for home use today. It was made of a galvanized iron water tank, painted black and set inside an insulated box covered with glass. By painting his tanks a dark color, Kemp enhanced their power as solar collectors; by insulating the box, he minimized conductive heat loss; and by covering the tanks with glass, he trapped the collected heat.

Kemp sold his patent to a pair of Californians named Brooks and Congers in 1895. These two Pasadena businessmen marketed the product to about a third of their fellow townfolk; then, along with a competitor called Day and Night, they entered the markets in Arizona, Florida, and the Hawaiian Islands.

The solar water heater was received enthusiastically, not as a radical means of harnessing energy, but rather as an efficient one. Its success in its first markets was due partly to the relative frequency of sunny days, and partly to the high cost of other fuels.

Early in the 20th century, coal was the principal fuel used in the United States. But California had little coal of its own; and imported coal cost twice what it cost elsewhere in the country. Natural gas production was still in its infancy, and gas cost about ten times what it costs today. Needless to say, electricity—which relies for its production on other fuels—was prohibitively expensive.

In this market, solar heating was relatively cheap, renewable, and easy to obtain. Other experiments were going on elsewhere in the world, including a 50 horsepower solar steam engine built near Cairo, Egypt, and a solar-distilling operation in Chile that produced 6000 gallons a day of fresh water from salt water. But through the 1920s and 1930s, the development of solar energy took place largely in the US—principally in Florida and California, and mainly concentrating on water heating. In the 1940s, however, the US was gearing up to participate in World War II, and prohibited the nonmilitary use of copper. Since plastics technology was relatively unsophisticated at that time, and only copper pipes could resist corrosion and be useful in a solar system, the progress of solar energy technology ground to a halt. By the end of the war, cheap alternative energies—gas, oil, and electricity —discouraged its further exploration, except as a topic of academic discussion.

In the 1980s, we find ourselves looking back on the earlier days of fossil fuels to see what we can learn from them. And solar energy is making a strong comeback. Throughout Europe, the market for solar hardware is booming. In West Germany in 1975, for instance, no companies were in that business; but by 1977, five hundred companies reckoned their sales at more than $20 million. The French were only slightly behind, projecting 200,000 homes and offices using solar energy by 1981. In the United States, the number of various solar systems in operation is approaching 500,000; and with increasing government support, solar installations are becoming more familiar and available.

The Department of Energy budget for 1981 established the Solar Energy Information Data Bank, whose purpose is "to collect, review, and disseminate information for all solar technologies." Solar application programs include a variety of incentives, such as tax credits for investments by residential users in solar energy equipment to heat or cool buildings. Direct subsidies for conservation loans to owners of residential buildings using solar energy are keeping some payments very low. There are also considerable incentives at the state level in almost every state. (For further information, see p. 10.)

Collection

You may think you don't know anything about solar energy collection, but you've probably already experienced its basic effects. If you have ever left your car in the sun with the windows rolled up for an hour or more, you returned to a car that was warmer inside than the air outside.

This happens because the sunlight passes through the car's windows to heat the interior parts of the car, such as the dashboard and seats, which in turn pass their heat to the air in the car. Because the windows are closed, the air is trapped and cannot escape so it becomes warmer and warmer.

Essentially, this is the basis for solar collection systems, which are constructed to absorb and trap as much of the sun's heat as possible.

Air outside the car heats up in the same way, but it is free to rise as it warms, and is replaced by cooler air from high above the ground.

Collection systems use the principle that dark colors absorb heat and light colors reflect it. You have experienced this principle first-hand if you've ever worn a light-colored shirt in summer, or a dark-colored coat in winter. Some sunny winter day when the snow is piled up, try this simple experiment: Put a piece of black cloth and a piece of white cloth next to each other on a snowbank. Soon the black cloth becomes wet—it is absorbing the sun's heat, and melting the snow beneath it. However, the white cloth stays dry—in fact, it may even insulate the snow against the sun and keep it from melting, depending on such factors as temperature and wind.

Solar energy collectors use this principle when they are painted or coated a dark color. A dark-colored collector will receive and absorb the sun's southerly rays. The combination of a dark-colored absorptive material and glass to trap the heat make up the basic elements of solar heat collectors.

The Greenhouse Effect

Sunlight is able to enter a greenhouse because it can penetrate glass. The sun's rays warm solid objects in the greenhouse, which radiate energy in their turn. But these objects radiate energy at a longer wavelength, which cannot penetrate the glass. So these rays are reflected and absorbed, and a good deal of their energy stays in the greenhouse.

For many years people believed that the ability of glass to trap the longer infrared waves was responsible for the heat gain in greenhouses during the day, so they called this phenomenon the "greenhouse effect." However, we now know that the greenhouse effect is responsible for only a small part of the heat trapped in a greenhouse. Most of the heat gain comes from keeping the warmed air in an enclosed space so that it is unable to escape.

If the greenhouse effect actually *were* the primary cause of heat gain in greenhouses, polyethylene greenhouses would remain cool in the sun, because the longer waves *can* penetrate polyethylene. But these greenhouses have the same heat gain as glass ones do.

However, the greenhouse effect *is* most important in heating the earth. Because our atmosphere is surrounded by a vacuum, and because convection currents and conduction cannot occur in a vacuum, the earth's atmosphere can lose heat *only* through radiation. If there were nothing to slow down the heat loss through radiation of infrared waves, the temperature of the atmosphere would drop dramatically, completely altering our climate.

However, water vapor in the atmosphere, like glass in the greenhouse, can be penetrated by infrared waves. The infrared waves are trapped in the atmosphere in just the same way that glass traps them in a greenhouse. The difference is that this trapped heat is crucial to the earth's atmosphere but not to the heat gain in a greenhouse.

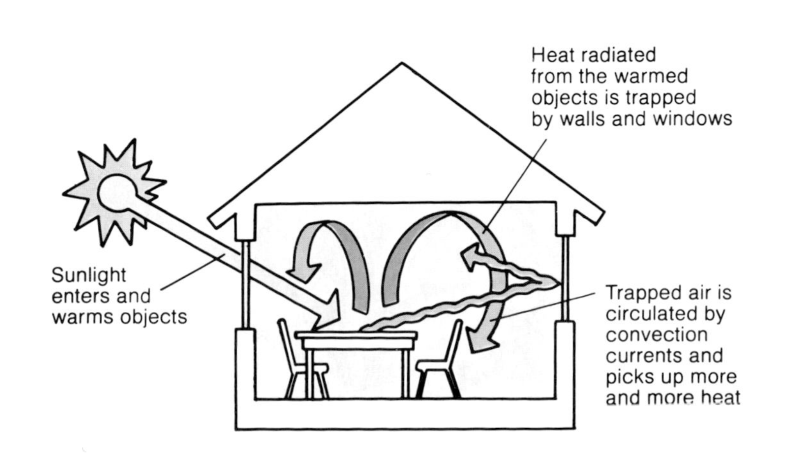

Storage

If solar heat is to be used when the sun is not shining, excess heat must be stored. A storage unit must have sufficient **mass** to store the collected heat. The dark cloth on the snowbank absorbed the sun's heat, but if you continued the experiment until sundown, you would see that it also cooled down rapidly. This is because it has very little mass in which to store the collected heat.

In some systems, called **storage heaters,** collection and storage take place in the same space. (The black cloth is an example of this principle.) In other systems, the collector is completely separate from the storage unit, and the collected heat is transported from one to the other.

In either case, the essential feature of a solar storage unit is the heat-retaining ability of the material it is made of. While the mass is subjected to direct sunlight, or while the air surrounding it is warmer than the mass itself, the mass will continue to absorb heat. But as soon as the surrounding air becomes cooler than the mass, it begins to lose its heat to those surroundings by conduction, convection, or radiation (see "How Heat Moves"). Therefore the most effective way of storing heat is in massive (that is, heavy and dense) objects that will retain the heat even in the absence of direct sunlight. Examples include an internal wall or floor made of masonry, brick, or stone, especially if painted a flat, dark color; dark-colored cylinders, drums, or tanks filled with water; thermal walls such as the Trombé wall and the tube wall (see pages 78-79); or bins of rocks.

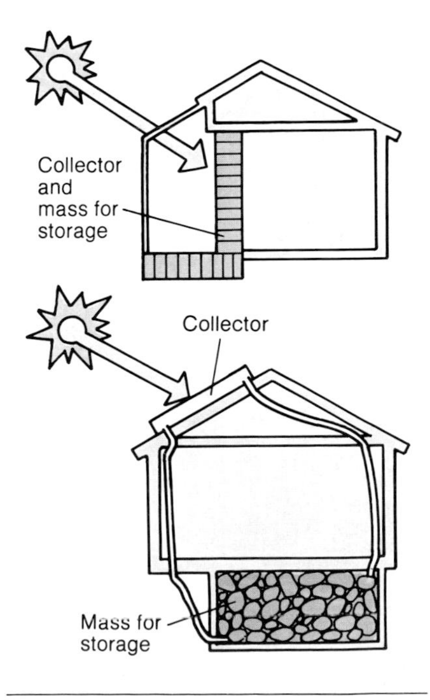

Distribution

Once the heat has been collected and stored, it must be distributed. For systems in which collection and storage are combined, the heat is distributed only from that point to wherever it is wanted. For systems in which collection and storage are separate, the heat must be distributed first from the collector to the storage unit and then to the desired location. The heat can be distributed immediately into the area surrounding the storage unit by the principles of natural heat movement, or by means of vents and ducts. Or the heat can be forced into places it would not naturally go, by means of pumps and fans. If the distribution is accomplished with the aid of architectural design but without mechanical devices, it is called a **passive** system. If mechanical devices such as fans or pumps are used to transport the heat, the system is said to be an **active** one.

Passive Distribution

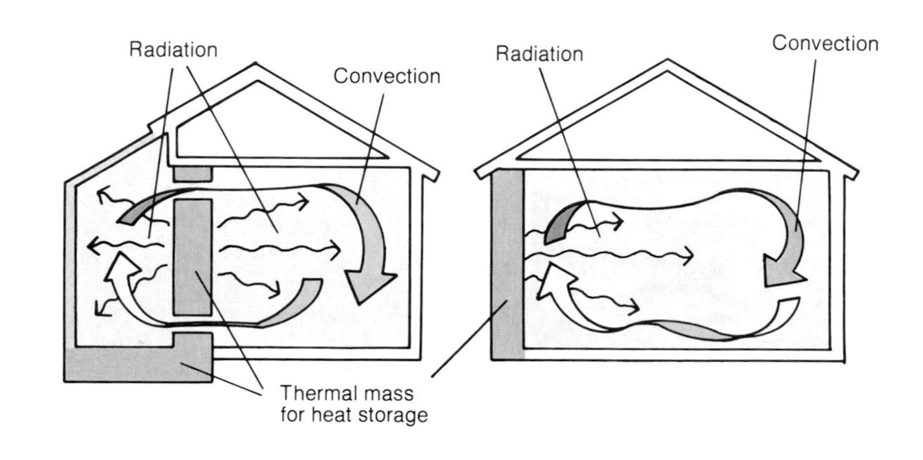

Active Distribution

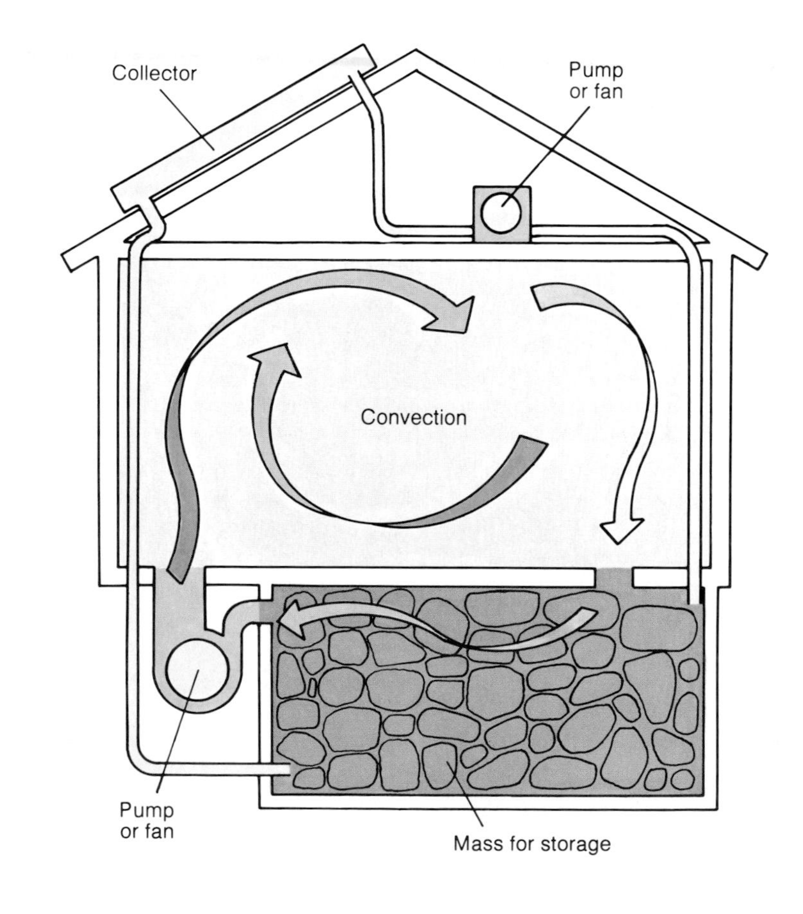

How Heat Moves

The natural movement of heat occurs in three different ways: conduction, convection, and radiation.

Conduction is the movement of heat through materials that are in direct physical contact, or within two parts of the same material. An iron pot, warmed by the fire beneath it, conducts heat to the stew cooking in it, to the sides and handle of the pan itself, and to the air immediately surrounding it.

Convection is the transfer of heat by the physical movement of a fluid (gas or liquid) medium which has been itself heated. Hot stew from the bottom of the pot rises to the surface to be replaced by the cooler surface stew. This movement is convection. The heat from the stew simmering in a pot then warms the surrounding air. The air itself rises after it is heated, and falls after it has given up its heat to the cooler ceiling, and is thereby cooled itself.

Radiation is the transfer of energy by ultraviolet, infrared, or visible waves. Ultraviolet and infrared waves are part of the light spectrum beyond what the human eye can see, and the infrared waves are primarily what we feel as heat. Infrared cameras work by photographic heat emanations; radiant heat is what you feel when you stand next to the stove flame, or next to a fireplace—one side of you is warmed, while the other side remains cool. Radiation is the only mode of heat transfer that can take place in a vacuum, and it is the way the earth receives heat through millions of miles of space from the sun.

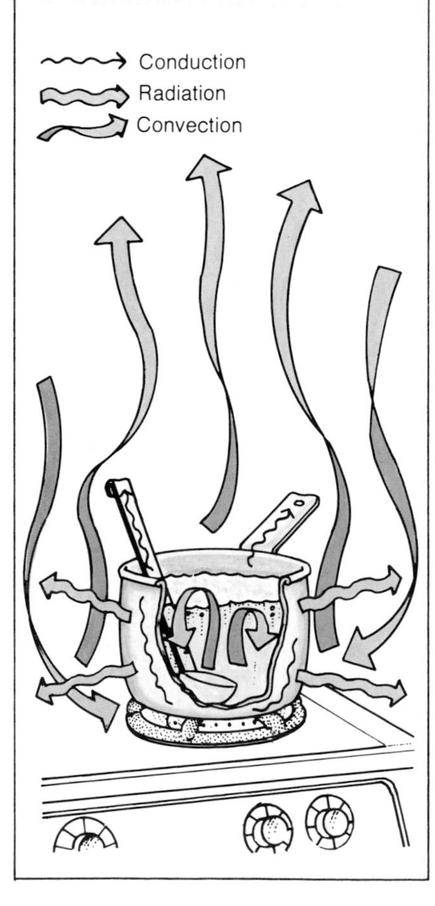

Passive Solar Heating Systems

Passive solar heating systems distribute solar heat as a natural extension of their collecting and storing properties. Such a system makes use of the sun's heat falling directly on a south-facing collector, which can be faced with glass or ultraviolet-resistant plastic. Passive solar systems use structural design to increase solar collection in winter and reduce it in summer. These systems use many south-facing windows, and have overhangs and awnings for shade in the hot summer months. The design of the house and placement of collection/storage elements are all-important.

■ **Direct gain systems:** Certain passive systems are said to be **direct gain systems** when they use the sun's heat without deflection or interference, and when collection, storage, and distribution all occur within the same space. In these systems, the space itself is heated directly by sunlight, and components of that space absorb, store, and distribute the heat. In general, such a system includes many south-facing windows, and a large mass within the space, placed to receive the most direct sunlight in cold weather and the least direct sunlight in hot weather. Typically, such a mass is made of a specific material that has the ability to collect, store, and distribute heat—for example, floors and/or walls of masonry, stone, or concrete, such as the adobe houses common to the American Southwest. Or sometimes this mass may be water, contained in drums or cylinders. The mass absorbs solar heat during its daytime exposure to direct sunlight, and reradiates that heat back into the space during the cooler night.

■ **Indirect gain systems:** Passive solar systems are said to be **indirect gain systems** when a thermal mass such as rock or contained liquid stands between the sun and the space to be heated. The mass absorbs the sun's heat and passes it on to the space.

An **indirect gain passive space-heating system** may be constructed in a room by adding a thermal mass several inches inside the south-facing window. That mass may be as simple as a collection of glass or plastic jars painted black and filled with water, or as sophisticated as a Trombé wall ducted at the top to permit warm air to pass into the room, and ducted at the bottom to permit cooler air to pass out of the room and back to the heat-collection area. (For more on Trombé wall, see page 78.)

You can also use rock, although its use is likely to be restricted to floors and walls. It's possible to bind piles of rock in place artistically with mesh wire; but for adequate passive solar space heating you would need about 100 pounds of small (1- to 4-inch) rocks per square foot of window with the sun shining directly onto the mass.

Thermosiphoning is simply a form of convection. When a wall (with or without drums or tubes for added heat storage) is placed directly behind a south-facing window, the heat is collected and trapped between the window glass and the wall. This heats the air there, which then rises, spilling into the room through vents at the top of the wall. The cooled air returns through vents at the bottom of the wall.

This principle can also be used with flat-plate collectors (see pages 80 to 82). Here, the elements normally associated with an active solar system are used in a passive system. In this case, collectors are always set below the storage tanks or bins to take advantage of the natural inclination of air and water to rise when heated and to fall when cooled.

Direct Gain

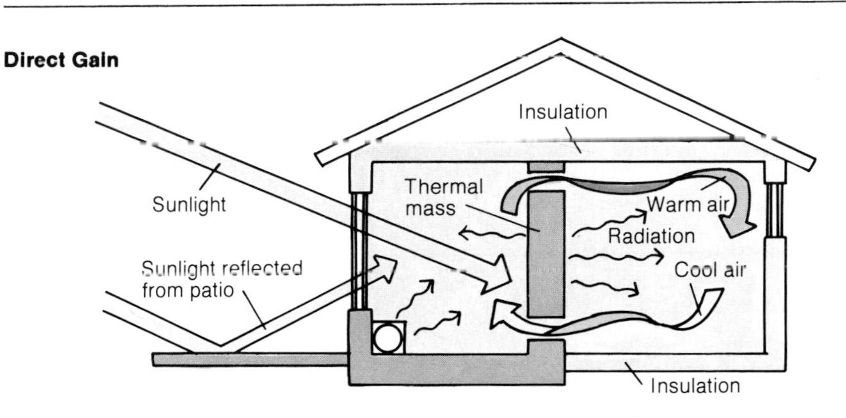

Sunlight passes through glass, and its heat is trapped in the room. Several thermal masses absorb the heat—the masonry floor, a stone window seat, and a stone wall. When the temperature inside the room becomes less than that of the thermal masses, they release the heat they stored earlier, keeping the room warm.

Indirect Gain

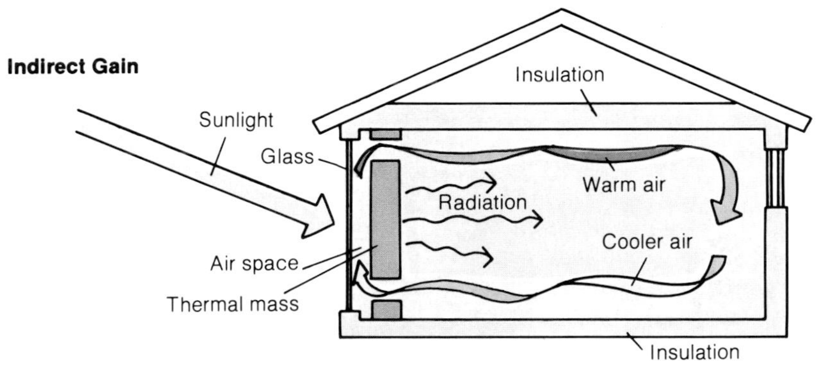

Sunlight passes through glass, and its heat is trapped in the narrow space between the window and the thick masonry wall. Vents at the top and bottom of the wall permit air to circulate and to heat the room by convection, and the thermal mass absorbs and stores heat for radiation into the room after the sun has gone.

Thermosiphon

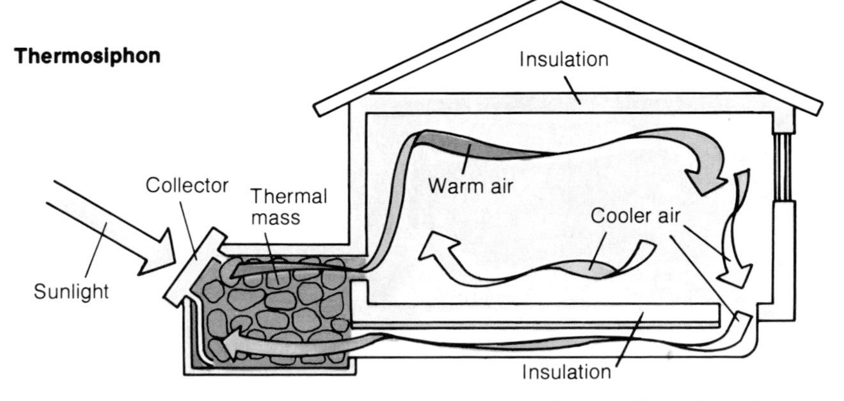

Sunlight is absorbed by the collector, and may or may not be stored by a thermal mass such as a rock bin. As the air in the collector or the rock bin warms up, it rises and enters the house through ducts provided for that purpose. As the air cools, it falls, entering return ducts that carry it back to be reheated.

Ways to Heat Your Space with the Sun

It's not necessary for you to invest in a complete, packaged solar system for thousands of dollars to take advantage of the sun's heat. Here are some ways to use the sun's heat without enormous outlays of money.

■ **Specially painted Venetian-type blinds:** Painted flat black on one side and shiny silver on the other, these blinds can serve two purposes. When the silver side faces the sun, it helps deflect the sun's heat—a cooling boon in hot summer weather. And when the black side faces the sun, the blinds themselves become solar collectors. By adjusting the angle of the slats, you can direct the heat collected by the black panels back into the room. A sophisticated version of this solar heat-collecting method installs the blinds between the two panels of a double-glazed window.

■ **Reflectors:** A reflector outside the window will reflect the sun's rays from the ground into your window. You can use what's already there—asphalt, or concrete (both will reflect between ¼ and ⅓ of the rays that strike it); or fresh snow (which will reflect nearly 90 per cent of the rays that strike it). Or you can improve on what you've got by buying or building a reflector made of tinfoil, Mylar, white porcelain, mirrors, white rock or sand—in fact, almost anything reflective. Just make sure that fragile materials won't be damaged by fencing them in, away from animals, children, etc. You can also use reflectors inside. (See the accompanying photos.)

A reflector can also be **part of an insulating wall** that's lowered by day to collect solar heat, and raised at night to insulate your living space.

■ **Masonry:** If you're willing to roll up the rug—permanently—you can create a direct-gain thermal mass by replacing your carpeting with floors made of **masonry tiles, stone, brick,** or **concrete.** These floors can be extremely attractive, as shown in the photograph on the facing page. The same principle applies to walls; but unless you are building your house from scratch, you probably will not relish replacing your walls.

■ **Greenhouse:** Add a greenhouse to the south wall of your house. This works especially well if the south wall already has a door or window. Both full greenhouses and windowbox greenhouses will provide indirect gain passive solar space heating. If you are interested in this aspect of solar heat, be sure to see Ortho's book, *How to Build & Use Greenhouses* which details the construction of a wide variety of greenhouses.

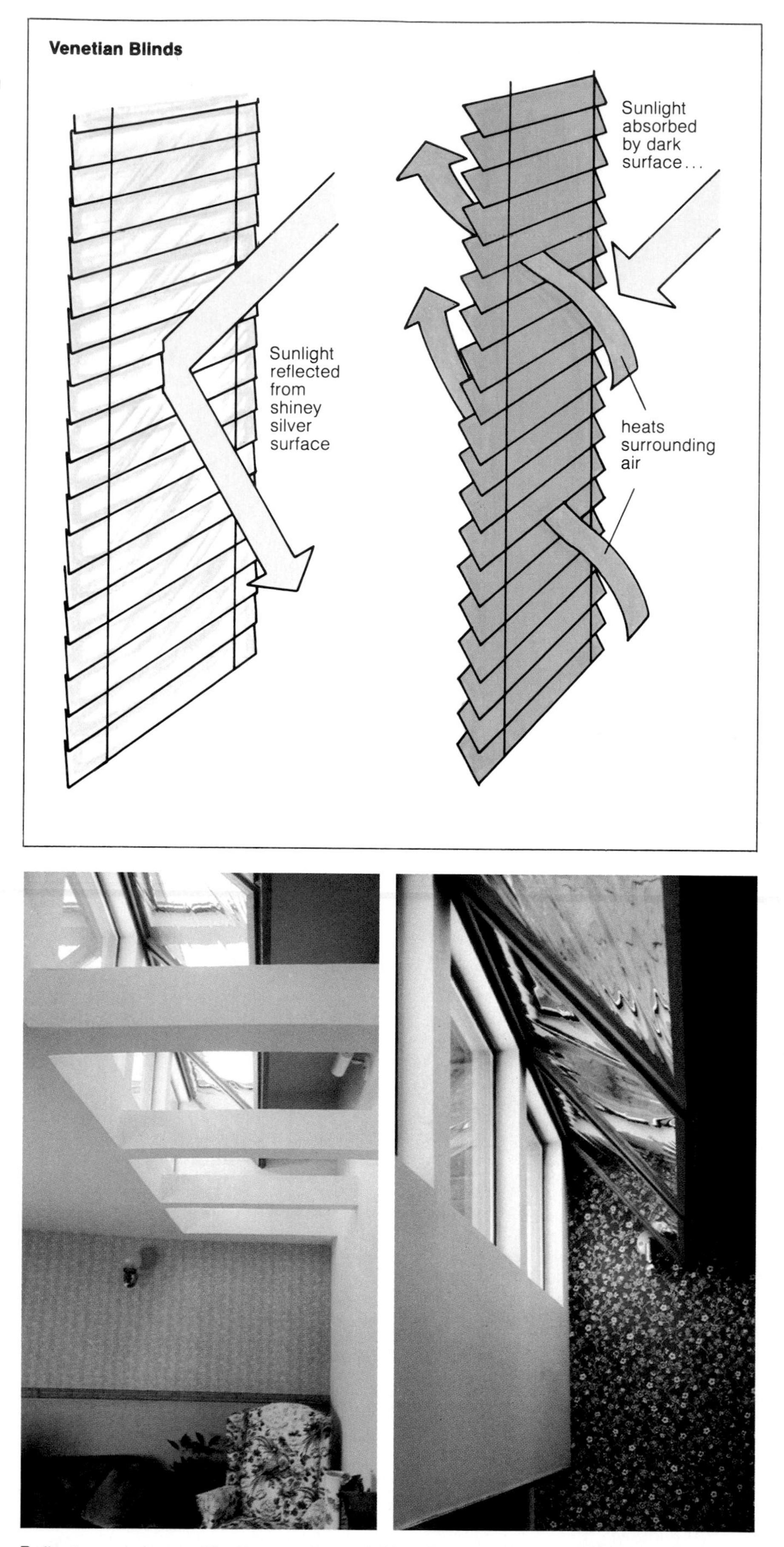

Reflectors can be used inside as well as outside to increase the amount of solar heat—and light—available to you. Make sure incoming light strikes your reflectors at an angle that will send the rays into your house, where you want them, instead of back out the window. Any kind of material that has a high reflective ability can be used for this purpose, although mirrors, porcelain, decorative metals, and metal fabrics are most commonly used.

Solar heating components can be beautiful as well as functional. The masonry floor in the photograph above (left), for instance, is a very well disguised heat sink—a stone mass which absorbs the heat from direct sunlight during the day, and stores it for nighttime use. The floor automatically radiates its stored heat when the air surrounding it cools below the temperature of the floor. The construction of the wooden walls (right) includes openings that act as vents or ducts to allow air to circulate within the house, carrying solar warmth throughout the house from the room in which it is collected, or from the masonry floor at night.

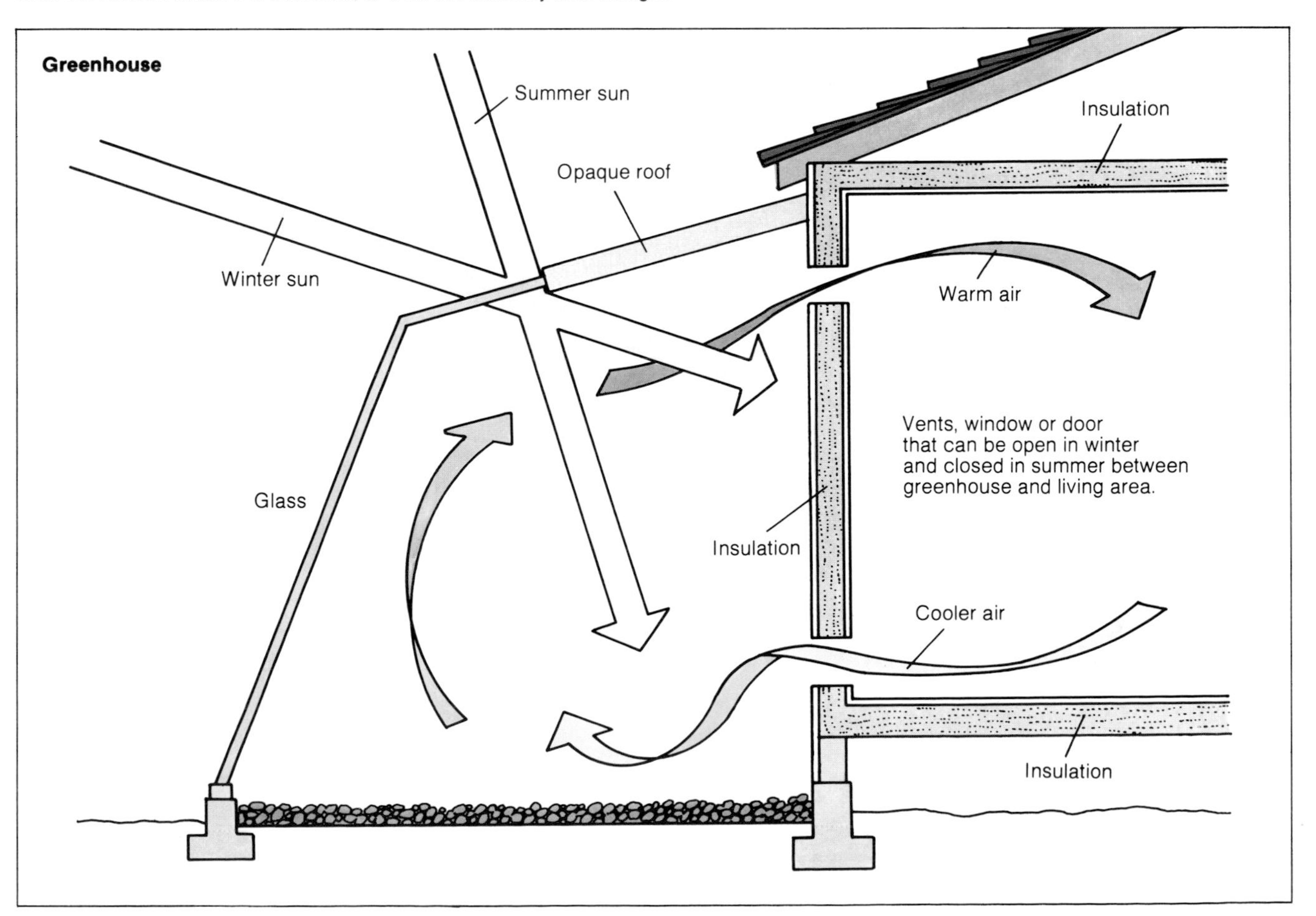

This sunspace is only a few feet wide, and was constructed expressly as a way of collecting solar heat. Heat is not only trapped inside the greenhouse-like space; it is also absorbed by and stored in the black-painted drums full of water. Heat can then be passed into the house by convection through a system of ducts at the top and bottom of the wall separating the sunspace from the living space of the house.

Careful engineering and thoughtful architecture allow the sun to do nearly all the work heating this house. Massive masonry walls act as heat sinks, absorbing and storing heat by day for reradiation to the building's interior at night. Three greenhouses trap the sun's heat for distribution by circulating fans and a system of ducts. The topmost greenhouse, which is the hottest, is used for growing cactus; the second has high masonry walls and is used for growing vegetables; the lowest greenhouse is used as a sitting room in which a variety of plants also grow.

The addition of a sliding roof over what had been open space adds both a room and a solar heat collector to another house. The frosted glass roof admits and traps the sun's heat; but because the roof slides open, excess heat can also be vented easily at any time.

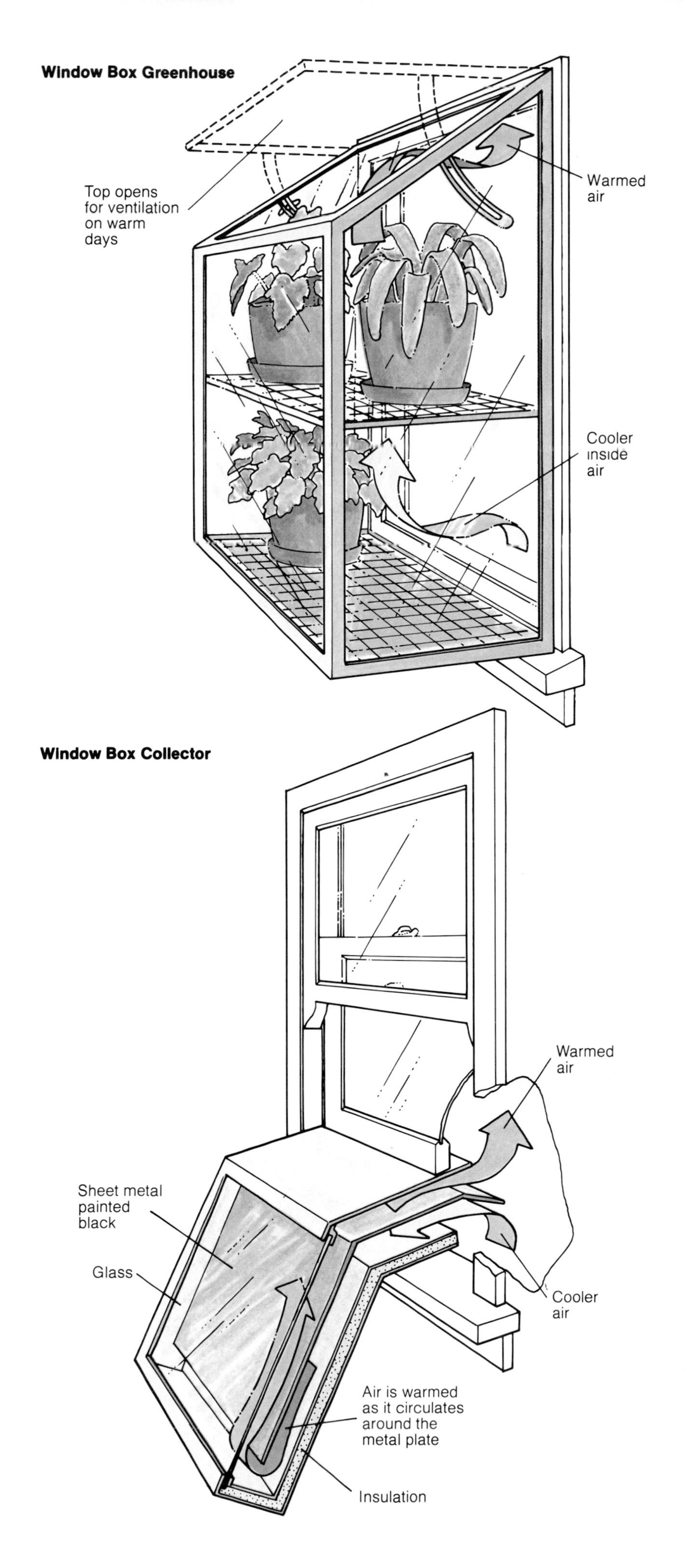

■ **Solarium or sunspace:** You can add a solarium, or sunspace, on one side of the outside walls. In principle, a sunspace is similar to a greenhouse—in fact, some people use it for that purpose. It is used to collect the excess heat brought in through doors, windows, or floor and ceiling vents. If you have a room with a good southfacing, suncatching window, adding a sunspace can augment the existing passive solar heating system. If you don't have a room with a good southern exposure, but do have one with a good east or west exposure and a little building space around it on which to develop a new southern exposure, that will work well also. Ideally, the entire south wall of the sunspace will become a collector.

For the floor, you can use a massive heat sink of some sort—masonry, stone, or concrete—which must be insulated from *underneath,* to keep maximum heat inside the sunspace. For additional heat storage, you can use rockbed storage or water-filled drums. If the wall between the potential sunspace and the house is currently an outside wall and you want to keep the rooms separate, be sure to insulate that wall. Make sure it has vents or ducts that are easy to open and close. A door between the rooms can provide additional heat transfer control.

■ **Drums or culvert pipes:** Paint a group of 55-gallon drums or culvert pipes black, set them in a south window, fill them with water, and you'll get a lot of solar heat for your effort. Units of contained water are easy to put in the window, in the path of direct sunlight. You can stack the drums on top of one another, or on their sides on racks. The drums themselves cost about $25 each from wholesale gasoline distributors although they are not usually treated to retard corrosion.

■ **Metal or plastic cylinders:** You can buy metal or plastic cylinders specifically designed to be filled with water, as you would the drums. Although the cylinders cost somewhat more than drums, they are more attractive and are not so susceptible to corrosion.

■ **Window-box collector:** Make or buy a window-box collector. It attaches to a south-facing window and pulls the cooler air from the floor through the collector, while the air passing over the absorber heats and rises. In other words, it uses the principles of thermosiphoning to provide indirect gain passive solar space heating. It will provide low-cost heating during the daytime only, since it can't store the heat at night. Therefore, you should be able to close it off with a small door or other seal, to keep the process from reversing itself at night. Thermosiphoning does not know indoors from out—it only knows movement from warmer to cooler.

■ Trombé wall: The Trombé wall was originally developed in 1956 by Jacques Michel, an architect, and Felix Trombé, a scientist. Despite variations since that time, its basic form hasn't really changed. A black-faced masonry or concrete south-facing wall, 12 to 16 inches thick, is used as the solar collector, but the wall itself is placed a few inches behind a sheet of glass. As the sun warms the air between the two surfaces, the air rises. At the top of the stone wall, ducts or vents allow the warmed air to pass into the room; at the bottom of the wall, ducts or vents allow cool air to return to be heated. All day, the natural thermosiphon process keeps warm air in the room; and because the heat takes from 6 to 8 hours to pass completely through this thickness of stone, the radiant heat from the wall itself warms the room at night. Dampers can be placed in the vents to prevent warm air from escaping through them at night.

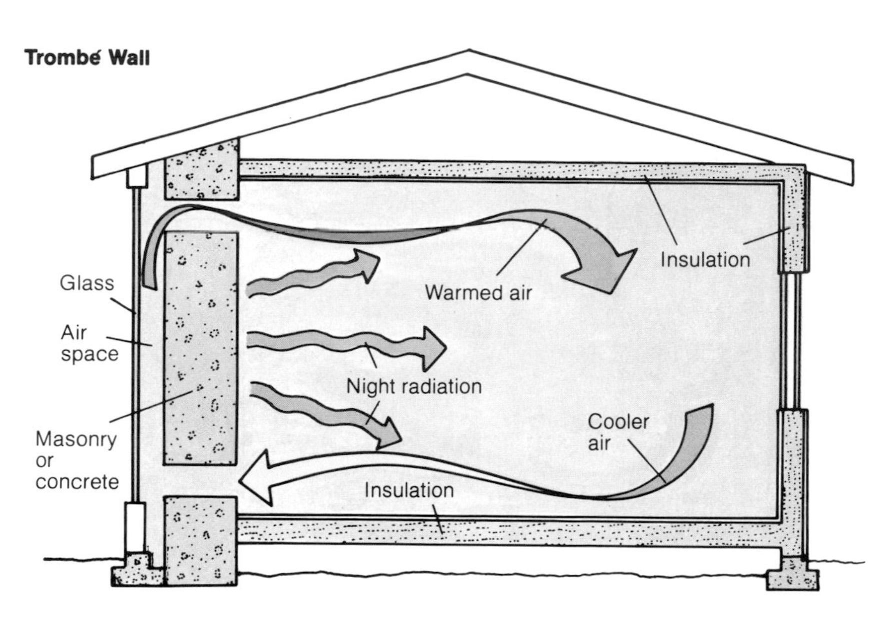

■ Tube wall: The tube wall resembles the Trombé wall, but instead of a masonry or concrete wall, a series of water-filled culvert pipe sections are placed vertically from floor to ceiling. A glazing on the south wall lets the sunlight pass through, warming the water in the tubes. At night, an insulated flap is placed between the water tubes and the outside air to radiate the heat into the room. The insulating flap can also be reflectorized to direct more of the sun's rays to the tubes during the day.

■ Roof pond: The roof pond is basically a large plastic waterbed mattress placed on a flat, dark, metal roof. During the day, the roof cover is opened to the sun; solar heat is collected by the dark roof itself and stored in the water bags. At night, insulated panels are closed over the water bags, allowing the absorbed heat to radiate down into the room.

The roof pond offers a natural solution to the problem of solar air conditioning—its whole process can simply be reversed in summer. The roof can be closed during the day, absorbing excess heat from the room below; and it can be opened at night to release its stored heat to the cool night air.

(Above) From the outside, the Trombé wall need not look very different from any other exterior wall. It should face south, in order to receive as much direct sunlight as possible. It absorbs the sun's heat for use warming the inside *(below)* both by radiating the warmth it has absorbed and stored in its mass, and by convection, when the warmed air rises and passes into the room through the open space at the top. Cooled air returns to the wall through an open space at the bottom of the wall.

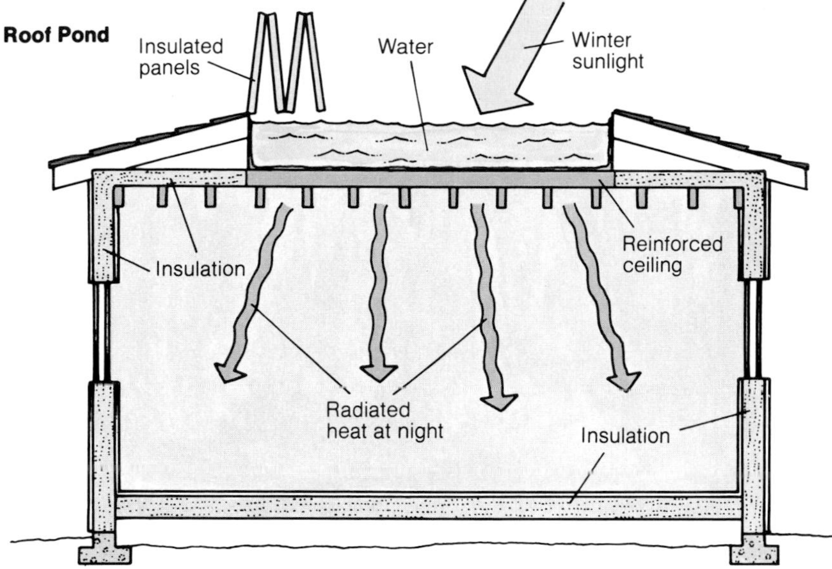

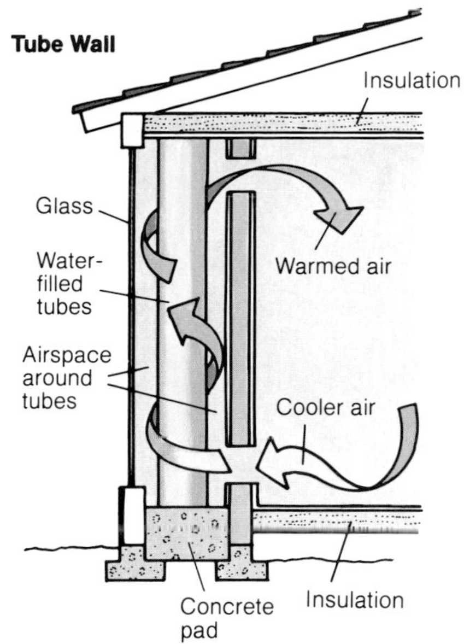

The tube wall is another way to use water for storing solar heat. Massive containers of water, placed in a southern wall, are exposed to the sun *(left and top)*. The heat they absorb is radiated to the interior of the house through the wall itself *(below)*, or else warms the living space by convection when operable vents are opened *(bottom)*. Particularly in cooler climates, it is important to be able to control the flow of air between the tube wall and the living space, since the water might become colder than the interior of the house at night, and would then begin to take heat away from the house, placing unnecessary demands on any conventional auxiliary heating system.

Active Solar Systems

As we saw earlier, active solar systems differ from passive solar systems in that mechanical devices such as pumps and fans are used to transport heat from collection to storage, or from storage to use.

Because active heating systems use mechanical components, they are generally somewhat more complicated than

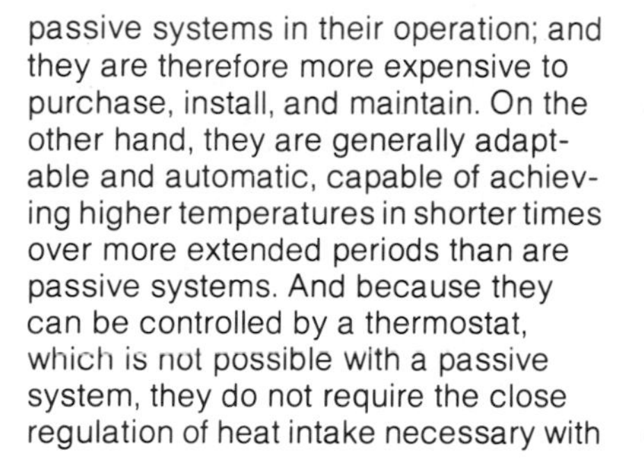

passive systems in their operation; and they are therefore more expensive to purchase, install, and maintain. On the other hand, they are generally adaptable and automatic, capable of achieving higher temperatures in shorter times over more extended periods than are passive systems. And because they can be controlled by a thermostat, which is not possible with a passive system, they do not require the close regulation of heat intake necessary with a passive system. Finally, active equipment is, on the whole, more compact and more readily adaptable to a variety of heating tasks than the cumbersome pieces of the passive package.

For all their evident differences, however, passive and active solar heating systems work on the same basic principles; and the success of any specific system reflects its ability to collect, store, and distribute the heat from solar energy in an efficient manner.

Air and Liquid Systems

Flat-plate collectors use either air or liquid (water, water-antifreeze solutions, or—rarely—oil solutions) to transport heat from the collection point to the storage unit, and then from storage unit to the desired points of distribution in the house.

If a **liquid system** (also called a water system or hydronic system) is used, the solar-heated water itself is transported to a hot-water storage tank, and carried from the tank through a pipe system to the space to be heated. A pump pipes the hot water from collector to storage and/or from storage to distribution.

If an **air system** is used, the solar-heated air is transported to a bin of rocks, which absorb and retain the heat. Air blown across the rocks picks up the heat and carries it through a system of ducts to the desired distribution point. A fan or blower is used to duct the solar-heated air from the collector to storage and/or from storage to the distribution point.

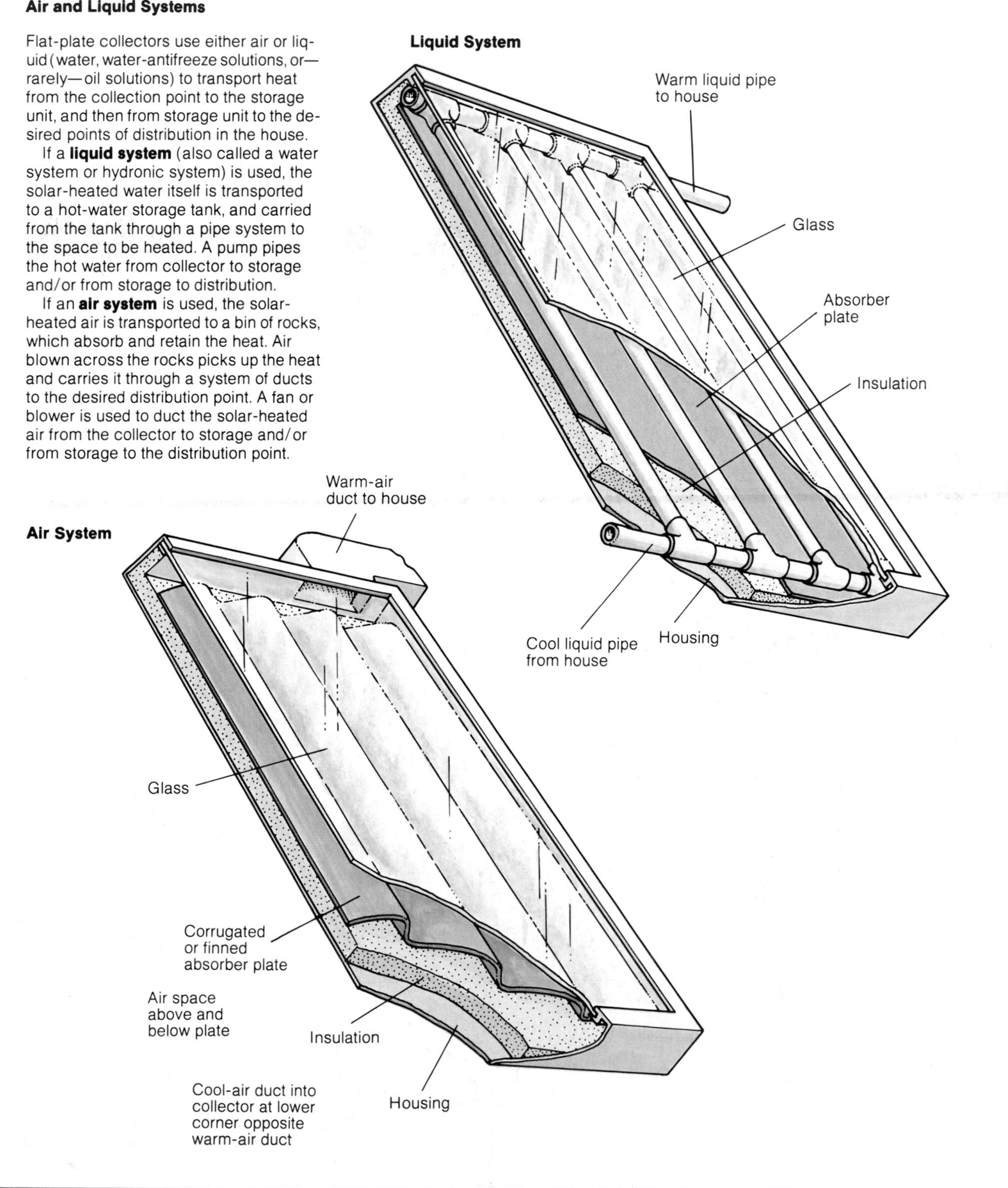

Collection

While the components of an active system are more compact than are passive components, they are also less likely to be integral parts of the structure itself; so the collectors, in particular, are by no means hidden. An active system will require a collector area equal to about ¼ to ⅓ of the square footage of the space to be heated. In other words, a 2000 square foot house will need approximately 500 to 700 square feet of collector to generate 75 percent of its heat, which is about the maximum heat most solar systems produce cost-effectively.

■ **Flat-plate collectors:** There are several kinds of solar heat collectors, of which the flat-plate collector is the simplest, least expensive, and most frequently used. Although flat-plate collectors come in several forms, they are all made up of an insulated box with a heat absorbing plate at the bottom, a heat transfer component (pipe or duct tubing) in the middle, and a clear cover at the top.

The absorber, which may be made of copper, aluminum, or some other material, is ordinarily coated black to enhance its heat collecting abilities. Or it may be coated instead with a chemical "selective coating" designed to absorb a high proportion of the available heat while losing little of it through emission.

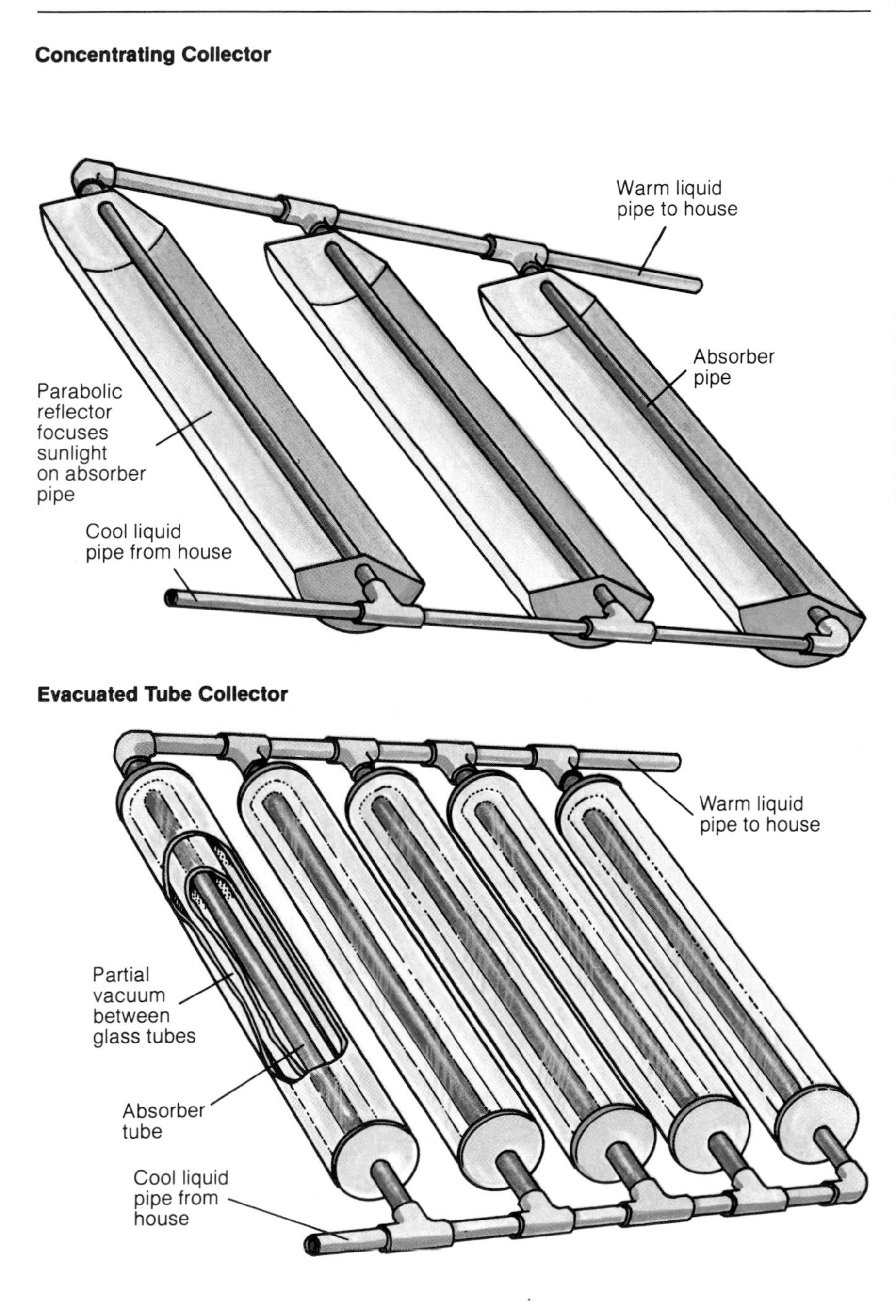

The primary purpose of the clear cover is to prevent heat loss from the collector by convection. Its secondary purpose is to keep the absorber plate clean. Most collector covers are made of tempered glass, although plastic and fiberglass are also used.

The cover should be removable, so it can be replaced if that becomes necessary; and it should be sealed. The seals keep the collector from leaking heat, as well as from gathering dust.

Thermostatically-controlled heaters will not turn on until they are warmer than the space they are designed to heat. Therefore inadequate insulation, which causes heat loss through the back and sides of the collector, is likely to require increased collection area. Since collectors cost about $10 to $15 per square foot *plus* installation, you have a vested interest in minimizing your collector's square footage requirements and seeing that its insulation is adequate. Usually it should equal at least the R-value of your wall insulation (see page 43); but if your collector is set on the ground instead of on the roof (see below), it may require an R-value as high as your ceiling's.

At present, flat-plate collectors, in all their variations, are the only *real* option in a completely active solar heating system. Compared with the other choices they are easy to use and readily available. Still, you might want to know of two other possibilities for the future, concentrating collectors and evacuated-tube collectors.

■ **Concentrating collectors:** These are made with mirrors and lenses, and ocus the heat they collect in the same way you would use a magnifying glass to start a fire. They can reach extremely hot temperatures—up to 1000° F. But because you are not likely to *use* 1000° F of heat in your home, such temperatures are very inefficient since in those ranges a high proportion of the heat generated is lost. Concentrating collectors also need direct sunlight—they cannot use reflected or otherwise diffused light—and must therefore be capable of tracking the sun. This ability, of course, can create a whole new set of mechanical problems for the owner.

■ **Evacuated-tube collectors:** These collectors are most effective at temperatures that far exceed anything needed to heat an ordinary house. If they ever become economically feasible for purposes other than industrial ones, they will be good house-sized solar heating components because their vacuum construction prevents them from losing heat in the cold or wind. This makes them reliably efficient in a way that flat-plate collectors (which do lose heat to the ambient atmosphere) are not.

Placing the Collector

Banks of flat-plate collectors are most commonly mounted on the roof or on a wall with a generally southern orientation. Such a setting keeps them away from the dangers of foot traffic, and off ground space that may be used in other ways. Some solar researchers suggest that collectors be mounted instead on racks on the ground for easy access, and to minimize the expense of improving the structure of your house to hold the added weight.

Where you place your collectors will be determined by a wide variety of factors. For example, if you need collectors only for your swimming pool, and it is not very close to your house, you may decide to mount them on racks. (See page 91 for rack-mounted collectors).

■ **Orientation:** A collector must be positioned to receive as much direct sunlight as it can. However, since the sun moves in an annual cycle as well as a daily one (rising farther north in summer than in winter, and riding lower in the sky in winter than in summer), and since flat-plate collectors are usually mounted in fixed positions, they cannot track the sun. Under such circumstances, facing due south is the most desirable position—in the US, the sun is always south of the ecliptic. However, morning haze and the relatively warmer afternoon temperatures make a slight southwesterly orientation even better. In fact, any orientation within 20 degrees either side of due south will give the collector enough sun to work efficiently, as long as it gets at least six hours of unobstructed access to whatever sun there is, every day, including December 21—the shortest day of the year.

■ **Tilt:** This is the angle at which the collector is aimed at the sky. The ideal tilt is your house's latitude plus 10 to 15 degrees for winter heating. Most commercially installed collectors tilt at 45 degrees, which about equals the sun's position in the North American sky.

If you are building your own system and want to be precise about orientation and tilt, remember that these numbers are generalizations, for use as guidelines. Specific situations demand specific solutions and you will need to explore the peculiarities of your own environment before you place your collectors. Speaking with an experienced solar architect or engineer, and with satisfied solar users, will give you information you'll never find in books.

Once your collector has gathered in all that heat, the storage unit is where it stays until you need it. For all practical purposes, three materials are currently regarded as successful for solar heat storage. Research continues into others.

Solar collectors can be placed virtually anywhere, as long as their exposure to sunlight is not impeded. When placed on the roof *(top)*, they are out of the way of foot traffic, and least likely to be shaded by nearby trees or buildings, but the roof's structural supports may have to be strengthened to accommodate their weight. If shadows from nearby objects do not create problems, you can simply place collectors out of the way where their orientation and tilt are optimum *(middle)*. Or they can be located at ground level *(bottom)* where winds that ordinarily carry away some of the collected heat are minimal, and where no added structural support for the house is necessary.

Concrete Water Storage Tank

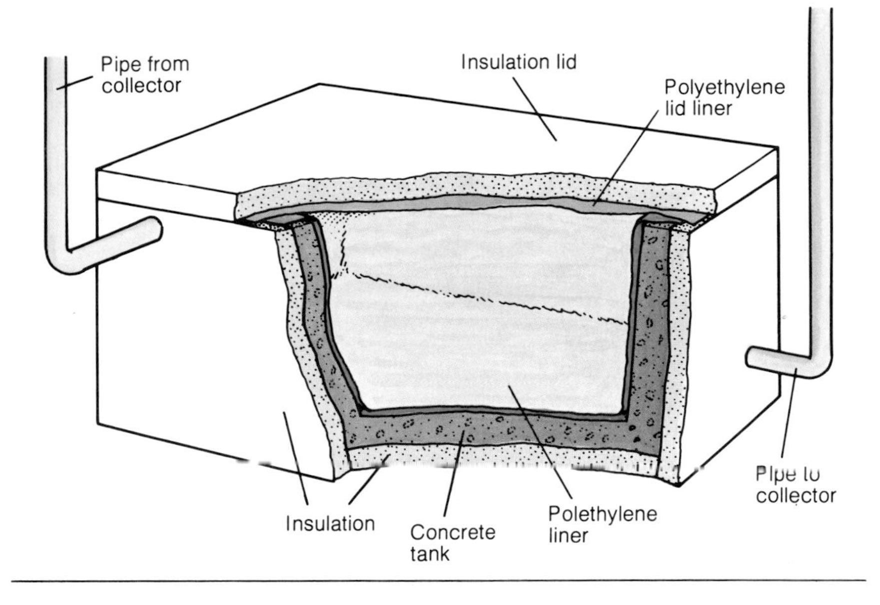

■ **Water:** Heat storage is usually accomplished with plain water or a water-glycol antifreeze mixture.

Water is inexpensive or free, and plentiful. It has the highest heat capacity per pound among the commonly used materials, and because its molecules flow freely among themselves, the heat it holds is distributed evenly by convection (see "How Heat Moves," page 72).

However, using water as a solar heat storage medium also has its disadvantages. Convection distributes the water's heat evenly, but it also evens out the temperatures throughout the storage tank. So as temperatures fall on one side, they also begin to fall on the other—and all through the middle as well. Therefore heat loss throughout the entire water tank occurs faster than in a rock bin.

The introduction of high-quality plastic waterproof liners has diminished the problem of leakage from poured concrete tanks, just as the use of glass tank lining has minimized problems with corrosion in domestic hot water systems. But these improvements have also led to an increase in the initial cost of water storage systems.

Solar collectors can also be set right into the ground if you have a hillside that is conveniently sloped to take advantage of the sun. Such placement is particularly usable for swimming pool solar heat collectors, because they have less need than other collectors to face directly into the sun, and their orientation and tilt need not be as specific.

■ **Tank size:** As a rule of thumb, experts say that you need anywhere from one to two gallons of water storage per square foot of collector to fill 75 percent of your heating needs; or that you need about one gallon of water per square foot of space to be heated; and so forth. Actually, there are many variables that dictate the size of a water tank suitable for your heating needs. One is the amount of heat you require, and this must be matched with the amount of heat that can be stored and the rate at which it is released. These calculations should be made by experienced solar architects and engineers.

Including a concrete shell, the filled tank can easily weigh five, ten, or fifteen tons. If you can build it in as an integral part of your basement, you have probably found the ideal location for the tank. If not, you will probably still place the tank in the basement, to prevent undue strain on your floorboards. If your basement is unheated, or if the tank abuts any exterior walls, it is particularly important to pay attention to proper insulation.

■ **Rock:** Like water, rock is cheap or free, and plentiful—although its abundance should not encourage you to go rock-picking as a do-it-yourself measure in building your rock bed storage bin. You can count on this bin occupying about three times the space a solar water storage tank would occupy. We are talking about a *lot* of rocks.

Cubic foot for cubic foot, water retains heat about twice as well as rock. But rock loses heat less quickly than water does, since heat doesn't circulate as directly through a pile of rocks as it does in a tank of water. Although one side of a rock bin may lose all its heat, the other will remain completely warm. This means that even if one side of the bin has cooled, you can still pick up heat from the other side. And unlike a water tank, nothing in a rock-bed storage bin will freeze, leak, rust, or corrode, nor is any part of the bin itself likely to wear out. Like a water tank, a rock bin must be insulated—and it must also be kept dry inside to discourage mildew. On the other hand, since rock gives up its heat more slowly than water, it is possible that the rate of your demand for heat may be greater than the rate at which the rocks give up their heat. The fact that water cools down faster also means that the heat stored there is given off faster and is, therefore, more available when demand is high. Again, the choice between rock and water will include all the variables of your particular situation, which should be discussed with an experienced solar expert.

Only air systems can make use of rock-bed heat storage. The cool air can simply be ducted or blown over the storage unit, picking up heat along the way, and then be sent on through the system to heat the house. Or the air may be allowed to circulate through the storage bin before being ducted out.

Rock-bin storage requires enormous volume, and the weight of a storage unit is impressive. Depending on all the variables, you may need 100 to 400 pounds of 1- to 4-inch rocks per square foot of collector area to maintain normal household temperatures. Whether you build your storage unit into a basement or into a crawlspace, it must be provided with a very secure and well-insulated foundation.

■ **Phase-change materials:** These heat-storage materials change their form as the temperature changes. The most common material is **eutectic salts.** These are extremely efficient heat-storage materials for their size and weight. They melt at low temperatures, and, at normal household temperatures, they absorb approximately two to four times as much heat as their weight in water, or up to twenty times their weight in rock. The smaller size considerably reduces the storage area needed, and the lighter weight allows you to use the materials in many locations, not just in a basement, crawlspace, or other heavily supported structure. To date, however, the problems with eutectic salts have made them impractical for solar heat storage.

Eutectic salts are prohibitively expensive in the necessary quantities, and their performance decreases markedly when temperatures rise or fall very much beyond the useful indoor range, making storage of fluctuating solar heat awkward. When they cool, they solidify only partly, causing a progressive breakdown over numerous phase-changes from solid to liquid and back to solid, which decreases their long-term efficiency.

Research is in progress which, it is hoped, will improve the stability of eutectic salts; but for the time being they are more on the leading edge of solar technology than in the practical mainstream.

Rock Storage

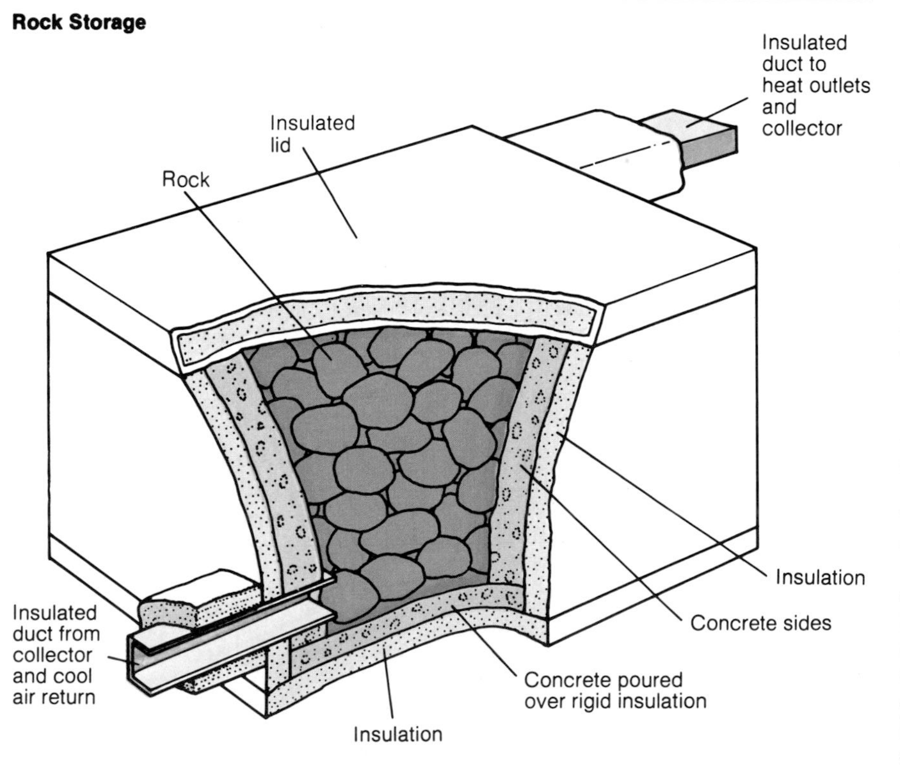

Distribution

Once the heat is collected and stored for later use, you still have to get it up into the house. Passive solar heating systems generally incorporate distribution as a natural extension of their collecting and storing properties. Active systems, however, are designed for greater user control. They demand pipes and perhaps heat exchangers for liquid systems, and ducts, fans, and blowers for air systems. And in almost every area of the United States, either system must connect with an auxiliary backup heating system.

Your auxiliary heating is your fail-safe device. Since solar heat cannot yet be relied on to fill all your heating needs, you want to know that you can stay warm when the cold winds blow under cloudy skies for a week at a time. If you live in a generally mild climate, your auxiliary heat might be a few electric space heaters. If you live where wood is plentiful and easy to bring home, an efficient stove or fireplace may provide your backup (see "Wood and Other Alternatives," page 95). But for most houses, auxiliary heating means the kind of full-bore conventional system that probably heats your house today—gas, oil, or electric. It can be operated separately from your solar system, making you responsible for turning it on when the time is right. Or your contractor should be able to connect the systems in such a fashion that all you need to do is maintain a constant thermostat setting; when your solar storage cools down from lack of sun, your auxiliary furnace will start up automatically.

In most cases, your existing ducts or pipes can be used to distribute solar heated air. Occasionally, you may prefer larger ducts because solar heat is usually lower than furnace heat, requiring the movement of more air to achieve the same level of comfort throughout the house. Further, depending on the degree of heat your system delivers, the current placement of heating ducts in your house may be less effective for solar heat than for your existing system. But for the most part, you can install solar components that can be connected to your existing distribution system.

This means that if you have a boiler, you are not required to install a water system. In fact, water systems demand particularly high collection and storage temperature to distribute their heat through baseboards, which can make the system somewhat inefficient. On the other hand, forced air systems operate at lower temperatures and can be more effective. You will want to examine all the alternatives for heat and cost-effectiveness to determine which will best meet your needs.

Air Heating System

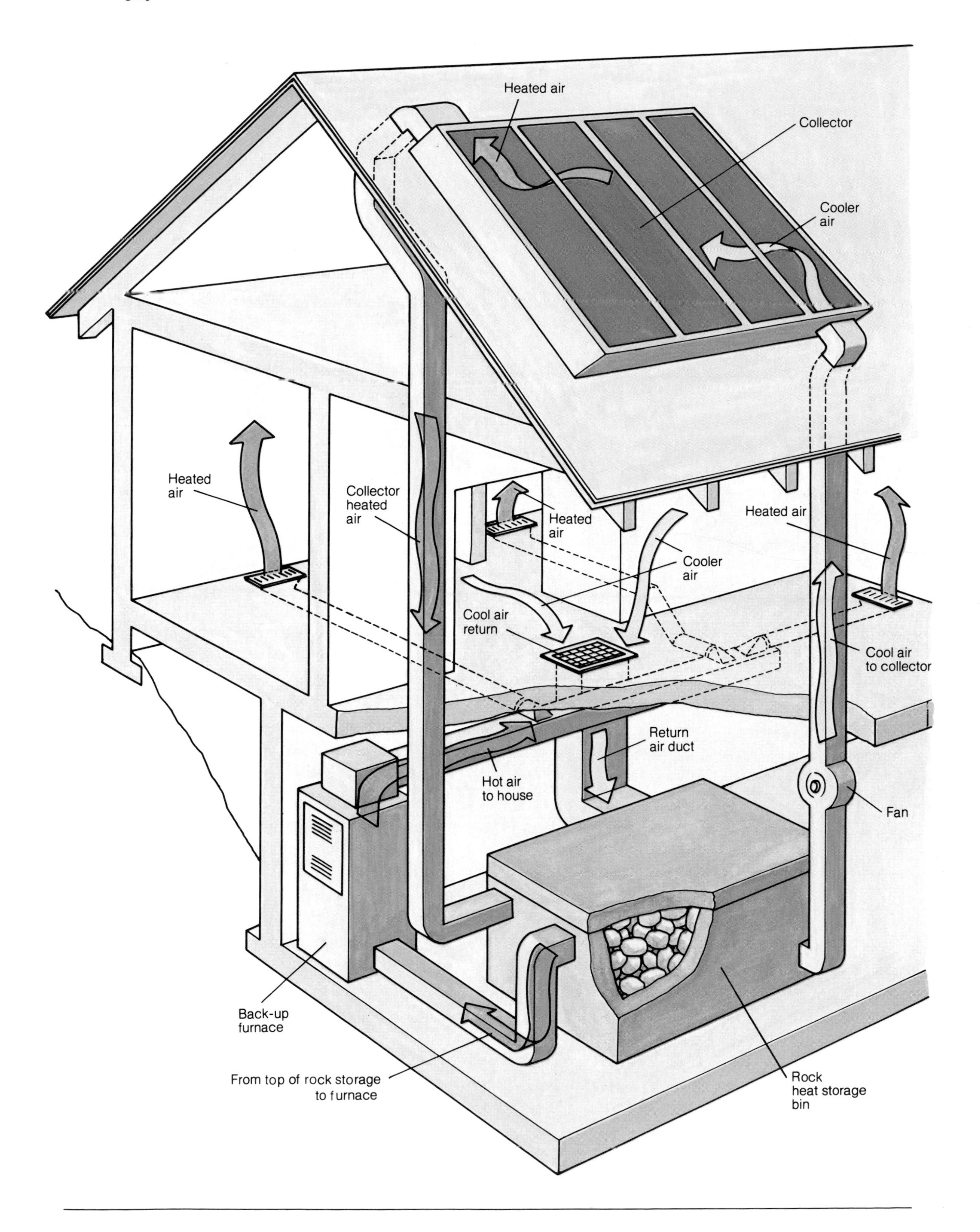

Water Heating System

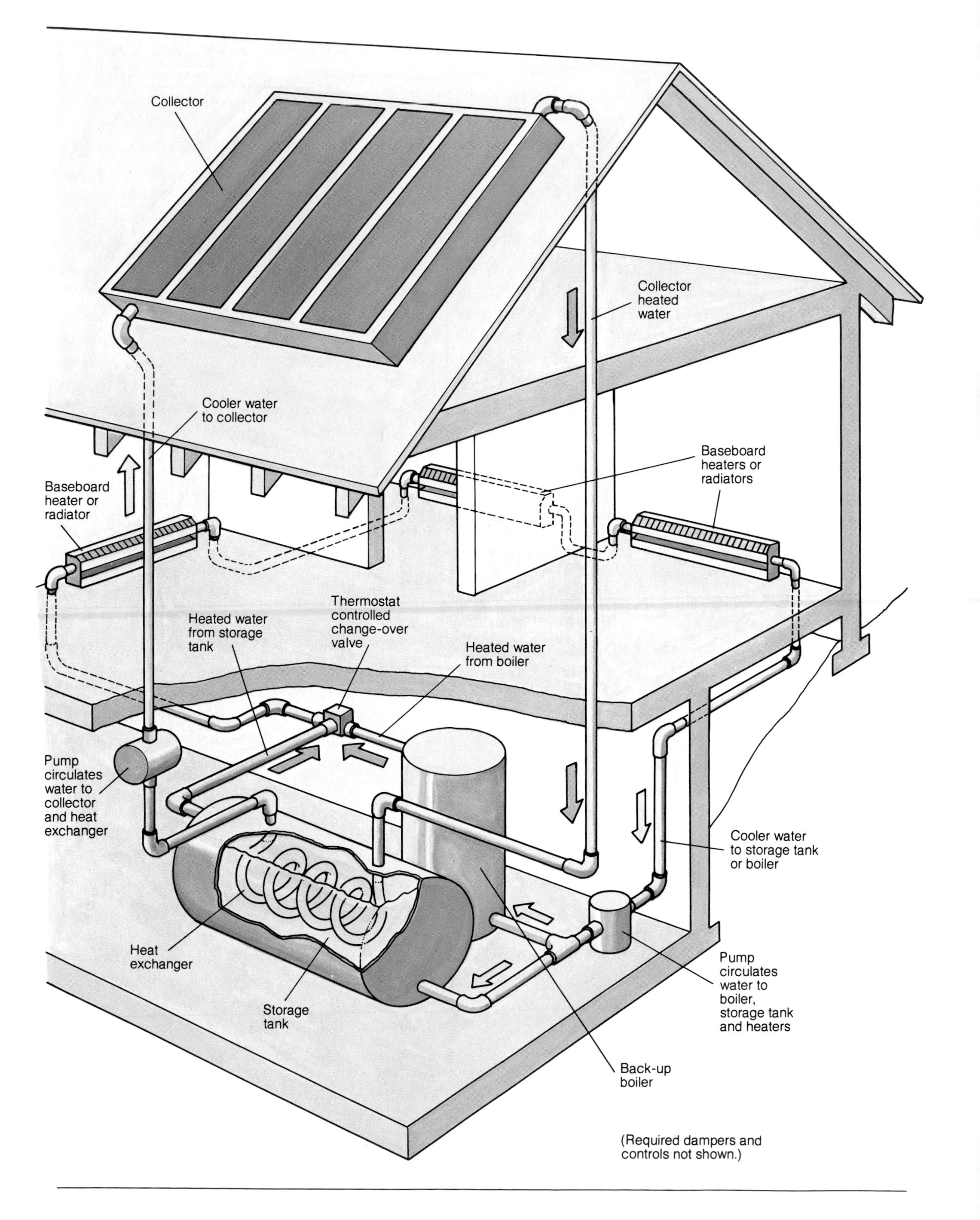

Domestic Hot Water

If you leave a black plastic bag full of water in the sun for a few hours, you will have a bagful of solar heated water whose temperature will range somewhere between pleasantly warm and much too hot to touch. If you add a hose with a shut-off valve to that bag, and suspend the whole unit higher than the outflow end of the hose, you can have solar heated water on demand, from the time the water warms up in the morning until early evening. If you add a second hose to the bag that introduces a fresh supply of unheated water as heated water is drawn off, you can use a slow but steady stream of solar heated water nearly all day.

If you want to be sure of a hot water supply all night as well as all day, you might simply wrap your water bag in a blanket of insulation every day in the late afternoon, and turn off the cold water influx at the same time. Or you might provide your system with an auxiliary water heater in case you're faced wth several days of heavy cloud cover in a row. If you want to use a lot of hot water at once from time to time (a lot meaning, here, more than your bag can hold), you can simply add an insulated storage tank, and pipe your heated water there by way of your outflow base. When you want it, you can demand hot water from the storage tank.

Despite our technological progress, the simplicity of the principles behind Clarence Kemp's 1891 water heater remain. And even at its most complex level, water heating is less expensive than space heating, and since hot water is used year-round, the cost is amortized more rapidly.

The average daily consumption of hot water in the United States is about 10 to 15 gallons per person. This use includes washing machines, dishwashers, and bathtubs, which consume a disproportionately high percentage of our hot water, and which operate principally in the morning and evening hours—before and after any non-storage solar hot water system can be effective.

As with all other solar energy adaptations for the house, conservation should be your first step because it requires the least effort or expenditure in proportion to the savings it returns. With solar heated water the savings are reflected directly and immediately in the size and complexity—and therefore in the cost—of the system you install. Less use translates into a need for smaller storage capacity and smaller collector area relative to the proportion of hot water supplied by the system. By the same token, greater use is reflected in a need for greater collection ability, and facilities for storing more heated water, or storing it for longer periods of time.

Bread Box Collector

The simplest form of solar water heater, such as the black plastic bag described above, is known as a storage heater, or **bread box** collector. In the bread box collector, as in the black plastic bag, the solar heat is collected and stored in the same unit. There is no transfer from one part of the system to another. However, all storage heaters lose at night the heat they collect by day. Storage heaters are essentially passive systems except at the use, or distribution, stage; an elementary mechanism allows user control.

The bread box has been around for half a century, in one form or another, but its basic design is always the same: one or more water tanks painted black and contained in an insulated box covered at the top (and sometimes on the south face as well) by double or triple glazing. The box is opened (usually by hand) in the morning to expose the tank(s) and glass to the sun; it is closed at night to minimize heat loss. The tank(s) can be filled by some independent method such as a hose, or connected to the general water supply of the house. The insulated solar heated water in the tank(s) is unlikely to freeze; however, if your area is subject to deep winters, you probably shouldn't use this system. In any case, insulate the pipes that carry water from or to the heater; this will guard against freezing and against ordinary heat loss.

With storage heaters, the collector/storage tank is ordinarily located above water outlets. This configuration keeps the system as simple as possible, and also keeps the cost down.

Bread Box Collector

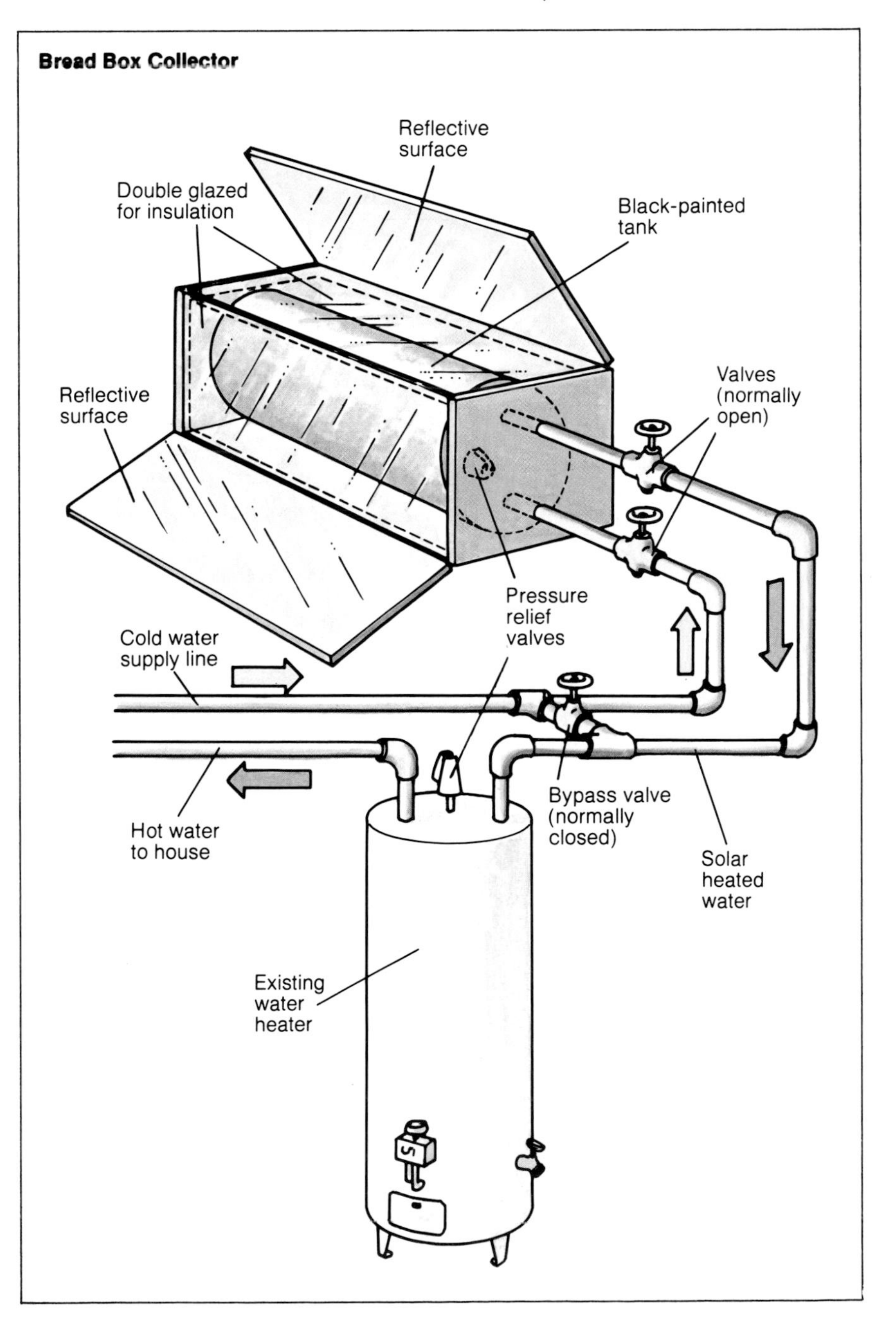

Thermosiphoning Systems

Solar heat can also be collected at a point below the storage tank, and transported by thermosiphoning to the higher storage. As mentioned earlier (page 73), thermosiphoning actually relies on gravity, taking advantage of the fact that warm air or liquid rises, and cool air or liquid falls.

In thermosiphoning systems, cool water circulates starting at the bottom of the collector (usually a flat-plate collector), is warmed, rises, and is piped into the *top* of the storage tank. As it cools, the water falls to the bottom of the tank and reenters the collector. Water for use in the house is tapped from the top, which is the warmest part of the storage tank.

The continuous cycle raises the temperature of solar heated water as long as the sun shines on the collector. How long the solar heated water remains warm after the sun ceases to shine on the collector depends on the capacity of the storage tank and how effectively the tank is insulated.

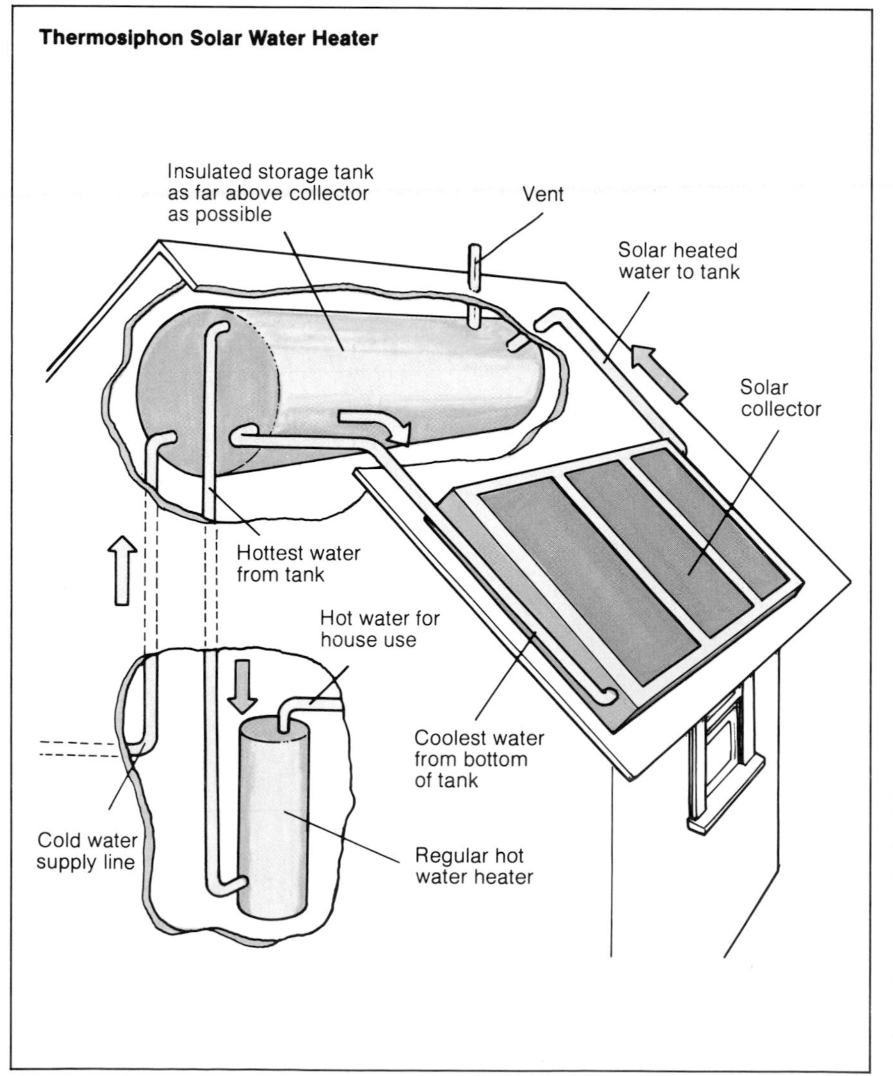

Active Hot Water Systems

Most solar hot water systems available on the market are not simply thermosiphon systems. They are, instead, active pump systems. As with active hydronic space heating systems, active solar water heaters use pumps to circulate water, and heat exchangers when antifreeze is used. They also are equipped with controllers that permit the pumps to work when the sun is capable of heating the water in the system.

The pumps most commonly used are 1/36 to 1/12 horsepower, depending on the number of collectors used, pipe size, and the height difference between the storage area and the collection surface.

The controls used to activate the system measure the temperature difference between the coldest part of the water storage and the absorber collection area. The pump is turned on and off by signals from the controller. In order for the pump to be activated, the absorber plate needs to be about 7°F warmer than the storage at its coolest point, which is indicated by sensors that are attached to both the collector and the tank. When there is a temperature difference, it means that there is heat to be gained on the collector surface, and the pump is turned on to retrieve it. As long as the temperature difference remains, the pump will stay on and the fluid will be sent to the collectors to continue the warming cycle.

When the sun sets in the evening, the fluid in the collector becomes cooler than that in the tank. The sensor again activates a signal, to shut off the pumps. The pump will remain off until the next time the collector is warmer than the coolest point of the storage water.

■ **Auxiliary system:** The auxiliary system is connected to your existing hot water system in the following way: Water arrives in your home from a central supply at a cool temperature. It is exposed to the sun through the collector and warmed. When hot water is required in the house, water is pulled from the hottest point of the storage tank into the domestic water heater. Since the heater's thermostat has been preset for the desired delivery temperature, the heater automatically "adjusts" the entering water to the temperature ready for delivery. For instance, if the water comes into your home at 60°F, and your desired temperature is 130°F, as soon as the solar system has heated the water to 100°F, about half your hot water heating has been accomplished by the solar energy system.

By setting the system up in this way, your home is supplied with hot water in virtually the same way as before, except that before the water reaches the standard water heater, the sun has preheated it, so that less work is required of your existing system.

■ **Insulation:** Good insulation on all pipes and tanks cannot be stressed too much. If there is any possibility that freezing or near-freezing temperatures are going to be encountered, insulation is crucial. It takes only one cold night to burst pipes that may require whole-scale replacement to repair your system.

One effective way to prevent your water heater's pipes from cracking in a freeze is to fill them with an antifreeze solution. Unfortunately, you cannot drink antifreeze and live to enjoy your solar hot water.

■ **Heat exchangers:** To get around this difficulty, **heat exchangers** are used. The most common forms of heat exchangers are copper or finned tubes immersed in the heated storage tank. A transfer fluid is pulled from the bottom of the coil, sent to the collector for heating, and returned through the top of the coil. As it travels down the tank, it "dumps" its heat into the domestic water supply.

There is also a "tube and shell" heat exchanger. This method uses two pumps: one to drive the heated fluid through one passage, and the other to send the water supply down the opposite passage in the other direction. Specific uses of heat exchangers may be regulated by your local building codes, which you should check carefully. But whatever you do, *do not* allow antifreeze into your household water.

■ **Water supply:** Although the optimum system for you will depend on your own particular circumstances, a useful rule of thumb in the United States is that a well-engineered system, installed and used properly, can supply a typical home with close to 100 percent of its summertime hot water requirements, about 25 to 30 percent of its winter requirements, and between 50 to 75 percent of its annual hot water needs. Costs for a hot water system will run in the vicinity of $2500, installed. Savings vary depending on use and the cost of your present water heating system.

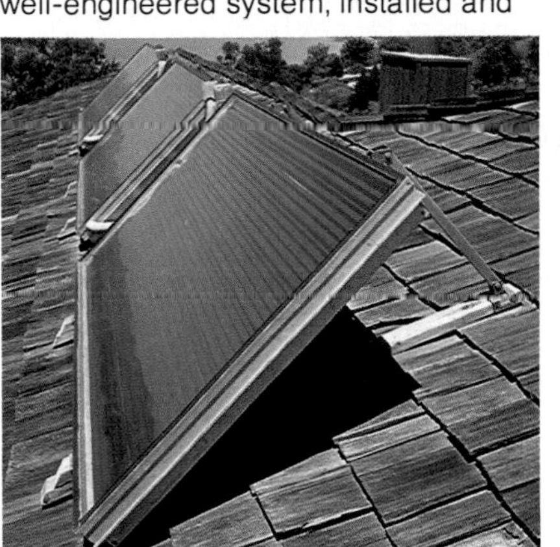

A solar water heater for domestic hot water consists primarily of two elements: collectors and the storage tank. In this house, the owners also installed a "chill-chaser", a small unit connected to the hot water tank with pipes and mounted on the end of a kitchen cabinet. It provides the added benefit of some space heat.

Residential Solar Water Heater: Comparison of Therms Used Before and After Installation

Month	Before Solar 1977	After Solar 1979	Therm Use Difference
January	189	192	+3
February	93	119	+26
March	106	70	-36
April	42	28	-14
May	54	12	-42
June	52	4	-48
July	52	2	-50
August	26	0	-26
September	24	0	-24
October	30	0	-30
November	84	48	-36
December	130	118	-12
Totals	882	593	-289 (fewer therms in 1979)

		Total therms
Jan. — April 1977	(before solar)	430
Jan. — April 1979	(after solar)	409
Jan. — April 1980	(after solar)	289

Closely kept records indicate how many therms of gas were saved after installation of system pictured at left. The more familiar the owners became with their system, the more they were able to save.

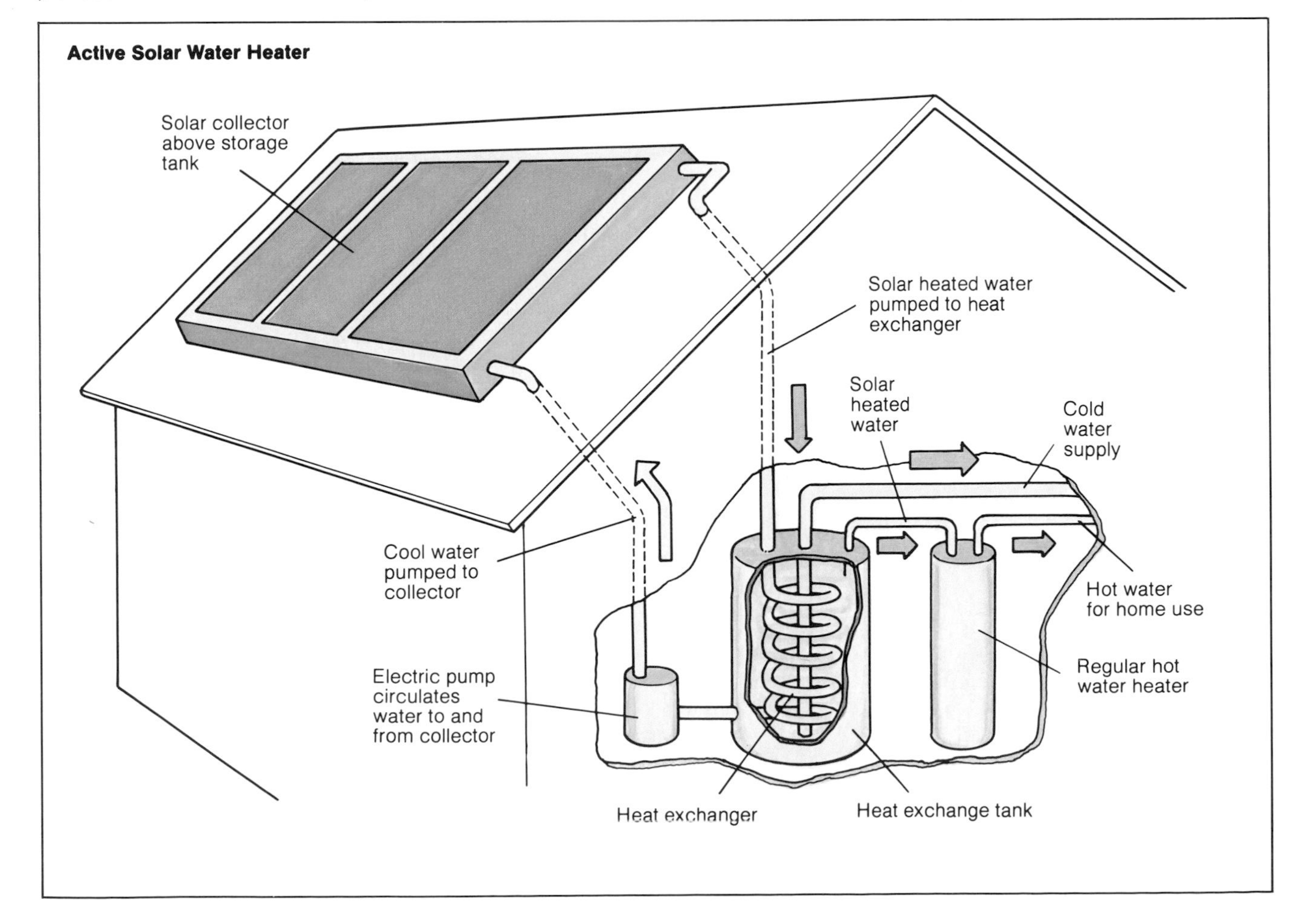

Solar Heated Swimming Pools

When you stop to think about it, solar energy seems to be a natural way to heat a swimming pool. After all, the water in the pool acts as a solar collector to some degree, absorbing about 70 to 80 percent of the available sunlight. However, this absorption doesn't necessarily bring the water up to an acceptable level of comfort, so you may want to enhance the water's natural collection ability by using an active solar system to heat it.

Conventionally, swimming pools are heated by natural gas or, less frequently, by electricity. The cost of heating a 16- by 36-foot pool to 80°F during the high-sun season (May to September), without any energy conservation measures, can equal the cost of driving a 20 mpg car 15 thousand miles. Even if you were willing to spend that much money providing all this heat through fossil fuel, an increasing number of local governments are restricting by law the use of fossil fuels for this purpose.

Solar heating offers a viable alternative; in fact, heating a pool with solar energy is one of the easiest uses of the whole solar technology. Why not use the sun to heat the vast expanse of water that swimming pools lay bare to the sun as well as to the air? Three characteristics of pools make them particularly adaptable to solar heating: (1) the water retains heat extremely well; (2) swimming pools are heated to relatively low temperatures; and (3) outdoor pools are likely to be heated only during high-sun seasons, in any case.

As with all other aspects of solar heating, the first step is to conserve what you already have. (See pages 66-67). Whether or not you use a conventional heating system, you can conserve whatever heat your pool absorbs from the sun by using a pool cover, particularly at night and on cloudy days. The same holds true when you have a solar pool heating system. There is no point in heating water that is cooled immediately by evaporation and radiation. Therefore, if you are considering a switch from a conventional system to a solar system, or thinking about installing a solar system alone, be sure to complete the picture with proper conservation measures.

An active solar swimming pool heater functions along the same lines as any other active solar water heater, except that the storage tank is also the point of distribution. A pump is used to convey water to be heated to the collectors, and to carry heated water from the collectors to the pool. As with other hot water systems, if there is any possibility of freezing, you must add antifreeze to the system, and use a heat exchanger to transfer heat from the solar-heated water to the water entering the pool. The simplest freeze protection for a seasonal solar pool heater is to drain the collector system during the winter.

An active liquid solar heating system is used to heat pools. All its parts should be compatible with your existing mechanism, since the pool water circulates both through your collector panels and through your present filter pump.

A swimming pool heater is a separate installation from the home's regular furnace. It is sometimes housed in its own small shed at the end of a mechanical chain that begins with a ½- to 1½-horsepower pump, and then continues with a filter system, the heater, a chlorinator, and finally the pool itself.

When you install solar pool heating, you break into the piping that runs between the filter and the heater; there a "pinch" or "gate" valve is installed, whose function is to redirect the circulating water to your solar collectors. The rerouted and solar-heated water returns to this break by way of a second pipe, and then moves on to the existing, thermostatically controlled heater. If your solar collectors have warmed the water sufficiently, it bypasses your heater. But if you have set your thermometer for, say, 80°F, and the water flowing from the collectors reaches the heater at 75°F, the heater kicks in. The operation is similar to any other solar heating household application that includes a thermostatically controlled backup system. Cool water flows from the bottom of the pool through the pump and filter, up or out to the collector panels, and back into the pool. Standard solar swimming pool heating systems also provide a heat sensor to start the system up and turn it off as the water's temperature fluctuates, and a vacuum relief valve to permit the system to drain, which minimizes their chances of freezing in winter.

Swimming Pool Sunlight Absorption

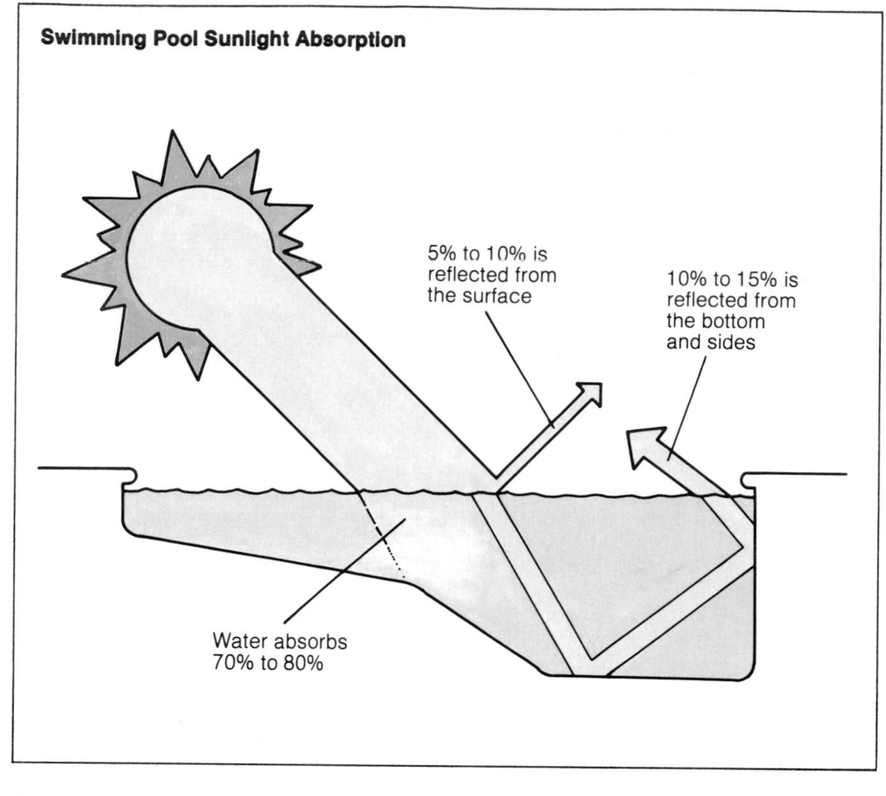

The important difference between solar-heated swimming pools and other active solar heating applications is the expense for the initial installation. In most locations, pool collectors do not have to be glazed. The most common kind of solar collector for pool heating is a 10- by 4-foot sheet of extruded plastic, including about 250 quarter-inch tubes lying side by side, in which the water can be warmed. The panels, wh ch cost about $10 to $15 per square foot installed, are light enough to handle fairly easily, and to install on most rooftops without added structural support. You should have at least enough collector area to equal 50 to 75 percent of the surface area of your pool. This will enable you to maintain a pool temperature of around 78°F. Because these panels are going to be used exclusively during the high-sun season, neither their orientation nor their tilt is as important as it is with solar space or domestic water heating.

Indoor pools can be solar-heated just as outdoor pools can be. However, they need only about half the collection area that an outdoor pool needs. This is because they are less liable to lose massive amounts of heat irretrievably—the convected heat loss warms the space around the pool instead of disappearing into the atmosphere.

(Left) A complete solar swimming pool heating system, before it is disguised by landscaping. The light, plastic collectors are up on racks; the circulating system stands below. *(Right)* A similar system, installed and in use. The collectors are fiberglass, and have been placed on the house to function *as* the roof, rather than being placed *on* the roof. In any case, solar swimming pool collectors are generally unglazed, and are, therefore, much lighter in weight than water or space heater collectors. As a result, almost any roof can stand up under their weight without additional structural supports.

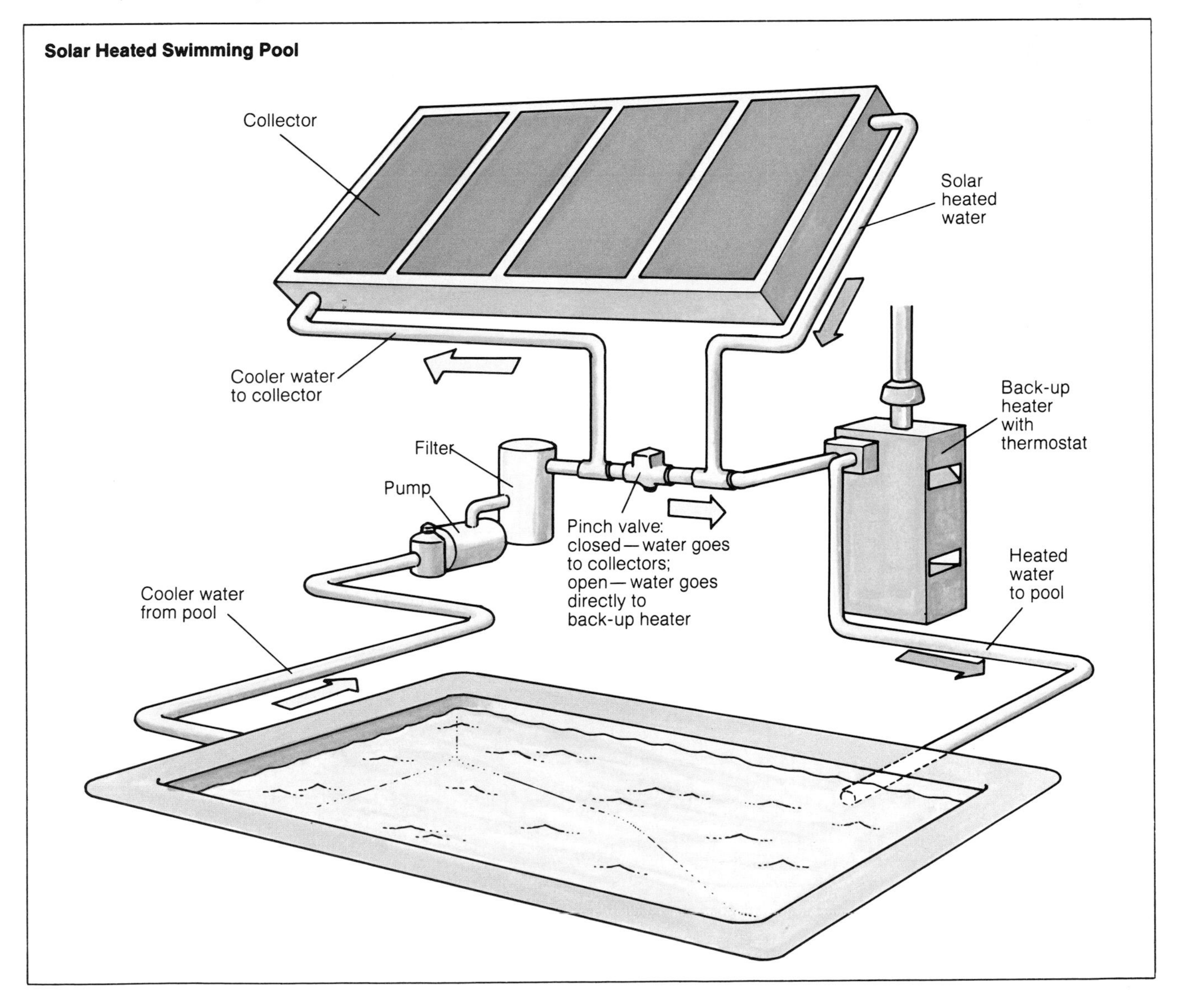

Is Solar for You?

Now that you have some idea of what kinds of solar systems are available and how they work, perhaps you find yourself squinting calculatingly into the sun, with an eye to using its energy to heat your water and/or living space. But before you rush out and purchase a flat-plate collector, storage tank, and pumps, and before you store your Bokhara rug in the attic in favor of masonry floor-tiles—indeed, before you even decide *which* solar system to use, you need to find out *whether* solar is really for you.

It makes no sense even to think about heating your living space with solar energy until you have thoroughly weatherized your house according to the steps already outlined in this book. Why go to the trouble of producing solar heat if it will only drift up through an uninsulated attic or out through the cracks around the windows? Thermostatically-controlled heaters will not turn on until they are warmer than the space they are designed to heat. Therefore, inadequate insulation is likely to require an increased collection area. And since collectors cost about $10 to $15 per square foot plus installation, it pays to minimize the square-footage requirements of your collector by properly insulating your house.

Once you're satisfied that your house can retain the solar heat you're thinking about tapping, then it's time to take a look at: your geography; your climate; financial considerations; and how you feel about solar energy.

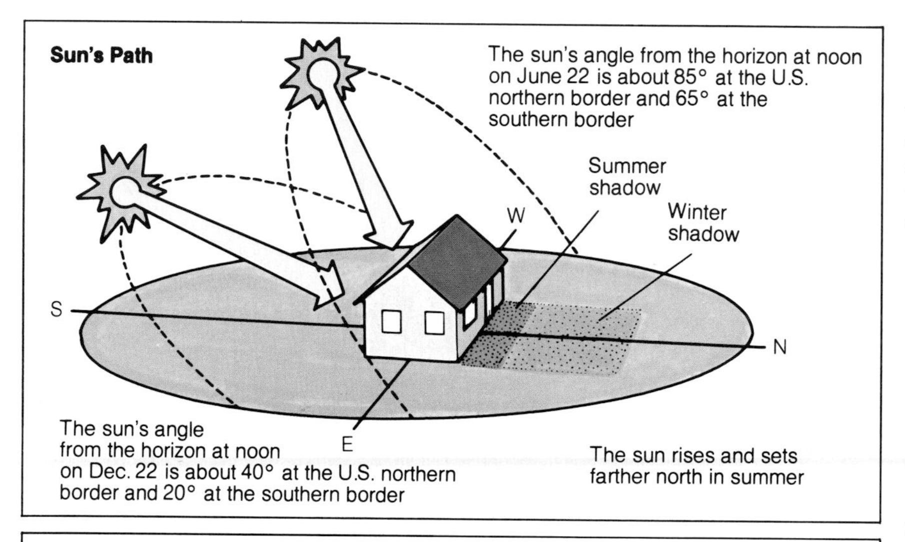

Solar Bands

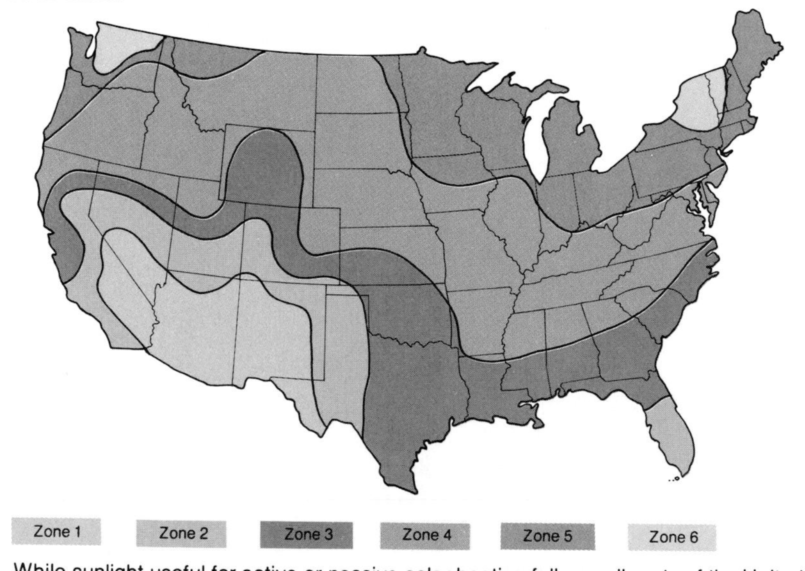

While sunlight useful for active or passive solar heating falls on all parts of the United States, more useful sunlight falls in some places than in others. The map above divides the continental United States into six principal sunfall regions, or zones. More useful sunlight falls in zone 1, where it is feasible that solar heating can accommodate practically all a home's annual heating needs, than falls in zone 6, where it is optimistic to think that solar heating can accomplish more than about 35 to 40 percent of the task. Useful sunfall does *not* bear a direct relationship to average daily temperatures.

Geography and Climate

Since your use of solar energy depends on how well you can exploit the sun's heat, the first task is to determine the relationship between the sun and your house. You need to find out how much you can take advantage of the sun and its movements.

■ **The sun's orbit:** The earth travels an elliptical orbit around the sun that brings the two bodies closest together (less than 90 million miles apart) on December 21, and carries them farthest apart (almost 96 million miles) on June 21. However, the earth is tilted on its axis at an angle of 23½° with the north pole toward the sun in summer and away from the sun in winter. This tilt makes the sun appear to be 47° higher in the sky in June than in December. Even though the earth is farther from the sun in summer, this angular shift causes the sun's rays to penetrate the earth's atmosphere more directly than they do in winter. As a result, our summers are warmer than our winters.

At the same time, the earth's daily rotation on its axis makes it appear that the sun rises in the east, passes overhead, and sets in the west. (Actually, it appears to rise and set slightly south of east and west in winter, and slightly north of east and west in summer; and it never passes directly overhead in the United States.)

■ **The orientation of your house to the sun:** More than any other factor, the position of your house relative to the sun will determine whether you can make use of solar energy.

In North America, most of the usable solar energy comes from the *south.* This means that you need to find out what kind(s) of southern exposure your house has or can have. Look for every place where you might collect and store solar energy, and ways to move the stored heat into your home.

Note everything that seems relevant to receiving (or not receiving) direct sunlight; also observe areas that do or could receive reflected sunlight.

Now get out a notebook and pen for making notes and drawings, and look at your house in terms of the following:

1. Does anything obstruct the sun's rays from falling on your house? (Look at the roof, the walls, and the southern face of the building.)

2. What is the general orientation of your house? (Note where the south/north axis falls in relation to your building.)

3. What openings are there on the south-facing surface?

4. What is the south-facing roof like?

Is it flat? pitched? Does it have vents, skylights, or anything else that would make mounting a collector difficult?

The next most important factor to determine is whether anything prevents the sun from reaching your potential collection area. In short, look for shading.

1. Are there trees to the south of your site that shade a potential collection point? Are you willing to prune them back or cut them down?.

2. Does a neighbor's house tower above yours, shading your south-facing windows?

Remember—you need access to the sun in order to use its energy.

You need a good idea of your particular solar microclimate as well as your regional solar climate. In general, every place in the United States receives enough direct and diffuse sunlight to make use of a solar energy system. Some parts of the country (for example, Miami Beach, Florida) get more sunlight on more days of the year than other places (for example, Seattle, Washington) and therefore the sun's heat is available for collection more often. But every part of the country receives sunlight. And the *availability* of sunlight—not the temperature—is what counts here. Any surface that can be exposed to sunlight can absorb the sun's radiant heat. As long as the surface's ability to collect heat, store the collected heat, and direct its flow is enhanced, even an extremely cold location will allow a solar energy system to provide home water and space heating.

In considering the potential effectiveness of a solar energy system, you have to look at your local weather and solar radiation patterns. How much solar energy is available on a clear day differs from one part of the United States to another, just as it differs from season to season, from day to day, and even from hour to hour. In general, the map to the left indicates areas of greater and lesser insolation (solar radiation received by the earth's surface). This will give you some idea of what your expectations can be for solar energy in your region. But if you find that your interest in solar heating is increasing, and you are seriously considering the possibility of installing some sort of solar system, you will want to locate more specific information for your area. There are many books which have elaborate tables and methods of calculation to help you determine the quantity of solar energy available in your particular region. Further, if you are new to the field of solar, you will want to seek out experts in your area who can help you determine the specific details relative to every aspect of a prospective solar system, including these initial calculations. Solar can be a real adventure, but you will want to be realistic.

The Financial Picture

Any building that is exposed to the sun is suitable for solar heating, to some degree. The rising costs of conventional fuels, together with a government policy that officially favors solar installation, encourage the exploration of solar heating.

Since 1974, the commercial market for active solar hardware has increased dramatically—thousands of homes per year are being at least partially adapted for solar systems in this country alone, and the government means to significantly increase that number by the end of the century.

As long as gas and oil were inexpensive the installation of a solar system could be a financially sound move only in a few places, and only for a few people. As recently as 1973, a payback period for a completely outfitted solar installation might have been as long as 25 to 30 years. By 1980, with tax incentives and rising fossil fuel prices, the payback can generally be expected to be 10 years or less. And if the prices of fossil fuels continue to rise at the rates projected by the Department of Energy, that payback period may fall to five years or fewer by 1990. However, while sunlight is free, the technology required to convert it into usable energy is not. You'll have to pay out something; exactly how much depends on the particulars of your house and heating needs. The best way to get a realistic idea of the cost is to have an expert do an on-site inspection. Then you'll know better what kind and size of solar installation is best for you; how much such an installation might cost; how much energy and money it may save you; and what its life-cycle costs and associated payback period are likely to be.

Because the costs for a solar system can range from $100 for a few 55-gallon oil drums to set in your window, to $15,000 or even more for a complete active installation, you should take the time for a thorough investigation of your options.

■ **Tax Incentives:** More than a million solar systems of one type or another are now in operation in the US. Government support is behind this switchover; the Solar Energy Information Data Bank has been established by the Department of Energy to "collect, review, and disseminate information for all solar technologies." To make solar use more attractive, the Federal government and many state governments offer tax credits or deductions (see page 9). These can be big or small carrots, depending on how much you end up spending on solar equipment, and on the level of your income.

The use of solar energy raises more than just a financial question—it brings up ecological concerns as well. In these pollution-conscious times, many people advocate solar energy because it is a clean, natural fuel. Others like the idea of being fairly independent of the fuelish foibles of civilization and want to try "producing" their own energy.

Whether your interest is ecological or romantic or a combination of both, you may be one of those people who feels that the satisfaction solar energy provides is as important as its ability to save you energy dollars. But if you are only interested in financial savings, you will want to make very careful calculations before making a major investment in solar equipment.

Now That You've Decided....

Your house is fully insulated. Your southern exposure will work. You've checked into the financial aspect and decided that you can swing it. Now it's time to seek out some expert advice on solar equipment.

Exploring the peculiarities of your own environment will help you find solutions to your own specific problems. Visit local residents who have already adapted their houses to solar energy. Speak with an experienced solar architect (for both passive and active installations), and with a solar engineer (for active installations). Check out local contractors who have adapted other houses in your area. Talk with them about their experiences. All of these people can tell you things you'll never find in books. They are certain to have some practical pointers that will save you time and money, whether they have sold solar equipment or bought it; whether they have installed solar equipment or lived with it; whether they like it or not. And because these people are local, they will know about your general climate and how that works with different systems. They may also be able to steer you to contractors, salespersons, and available hardware and alternatives, as well as to a variety of installed systems you can inspect to see what most closely suits your own needs.

Some solar equipment is easy to install; these jobs you can probably do yourself. Others, however, may be quite complicated and require the services of a trained specialist. If you do have to hire a contractor, be sure to reread page 49 "How to Choose a Contractor." The principles that apply to hiring a contractor to install foam-in-place insulation are identical to those that apply to hiring someone to install your bank of flat-plate collectors.

Fuelwood is readily available, easy to use, and provides efficient heat when burned properly, particularly in a wood-burning stove. Wind and water are also available as energy sources, although they may be expensive.

Wood and Other Alternatives

As late as the mid-1800s, wood provided about 90 percent of all the fuel energy consumed in the United States. But by the early 1900s, the serious exploitation of coal had brought this figure down to about 45 percent; and by 1970, the equally serious exploitation of oil and natural gas had reduced it even further, to about 1.5 percent. (However, wood accounted for about 8 percent of Sweden's fuel and 15 percent of Finland's during the same year.) In 1973, fewer than 1 percent of the homes in New England—a heavily-forested area with a long tradition of wood burning—used wood as the principal heating fuel.

However, since the Arab oil embargo in 1973, energy consciousness has risen along with the price of fossil fuels. Today, about a fifth of all New England houses are heated primarily with wood, and about half of all houses in this area use wood at least for auxiliary heating. Even though it isn't cheap to buy a good wood-burning stove or to turn a decorative fireplace into an efficient heat-producer, wood is again becoming a popular way to heat your house. Wood decreases your dependence on other forms of energy, and at the same time it lets you carry on one of America's more self-reliant and romantic traditions.

Until recently, the act of burning wood to produce heat incited anxiety and controversy among environmentalists and conservationists. The great fear was that it would not only pollute the air but, worse, utterly destroy a vital natural resource. However, the science of forestry management, combined with technological advances in the mechanics of burning wood, has shown that if the fuelwood available in the United States is properly managed, harvested, cut, seasoned, and burned, it can provide about 35 million homes with 100 percent of their heating fuel annually. That's more than half the residential heating requirement of the entire nation. (*Fuel*wood is distinct from lumber, paper and product wood, and the wood that must rot in the ground to maintain the soil's nutrient balance.)

Further, these figures don't take into account such factors as roadside, backyard, and other randomly available trees; orchards; branches and deadwood from harvested trees; or the trees standing on about 250 million acres of marginal forest and parkland.

Pollution issues from the inefficient use of wood as a fuel. The combustion that we call "burning" is essentially a speeded-up version of the oxidation that takes place naturally when a tree dies and decays. Decaying trees give off certain chemicals that might be harmful, but these are mitigated by the soil and surrounding flora. When wood is burned incompletely, these chemicals—essentially the same ones that forests have given off for eons —become part of the smoke that escapes up your chimney. The long-term effects of the chemicals aren't known, but they are presumed to be small.

However, when wood is burned *completely*, the pollution issue goes up in smoke: most of what is given off is water vapor, carbon dioxide, and heat. The efficient burning of wood creates the least pollution, as well as the most possible heat.

You may not be destined to manage or harvest large forests, but unless you live in the desert (and even then, there's mesquite!), the next few years will probably offer you the chance to cut, season, or at least burn wood.

This chapter will tell you what you need to know about wood, and about the increasingly popular wood-burning stoves so that if you choose to use it as either a primary or an auxiliary fuel, you can do so in the most efficient and pleasurable way possible.

Wood As Fuel

Because wood is readily available, inexpensive to use, and easy to integrate into other heating systems, it is an attractive heating fuel. Almost anyone can use it for auxiliary heating, and people who live near a source of wood may choose it as their primary fuel, especially if they are willing to undertake the rigorous and sometimes tedious effort of obtaining it themselves. In fact, about two-thirds of the wood used to heat American homes this year will be cut by people who intend to burn it themselves.

However, as a fuel, wood does have a few problems. It is bulky, and even the densest varieties occupy about five times the space taken up by oil for equivalent heating value. Wood is also relatively inefficient by weight: pound for pound, it provides only about half the heat produced by oil, and a third of the heat produced by coal. Its bulk and weight make transportation costly and storage awkward. While these problems make wood an unlikely candidate for solving our nation's energy problems, wood can be an energy-efficient fuel for some households.

■ **How wood is sold:** Wood for home heating is ordinarily sold by the **cord**. A cord is a pile of wood eight feet long, four feet high, and four feet deep. Its total volume, therefore, is 128 cubic feet. Of this, about 48 cubic feet are assumed to be air between lengths of wood, leaving 80 cubic feet of solid wood with an average moisture content of 20 percent. Wood may also be sold by the **short cord** (or **face cord**), measuring eight feet long, four feet high, and *any* depth—although it's usually between one and two feet, or the depth of one log, as cut.

Depending on the quality of the wood you buy, the time of year you buy it, and the distance it had to travel to get to you, you will pay somewhere between $30 and $200 for a standard cord of wood. For convenience, you can buy heating or fireplace wood in small packets from your local hardware or grocery store; these are easy to carry home and set into the fireplace. However, this convenience item will cost the equivalent of $600 to $700 per cord; so if you plan to use any large quantity of wood, buy it in bulk.

■ **Dry versus wet wood:** If you do purchase wood by the cord, it's best to buy dry, seasoned wood rather than wet, green wood. This has little to do with saving money—wood swells very little when wet, and a volume of wet wood doesn't contain appreciably less fuel energy than a similar volume of dry wood. Nor will dry wood give you significantly more heat than wet wood, although it will produce that heat more efficiently, since it doesn't have to burn off moisture and thus won't waste heat.

No, the advantage of dry wood is that it weighs noticeably less than green wood. This makes it easier for you to lift it, both in bulk and by the individual log. Also, since dry wood lights faster than wet wood, you need less kindling to start a fire. Other reasons to buy your wood dry rather than wet are that it burns more readily and stays lit more easily; it is generally cleaner and easier to handle; and it produces less creosote (a problem we'll discuss when we talk about chimneys on page 107).

Whether your wood is wet or dry, however, the mere size of your woodpile won't tell you how much potential heat you have stored up for the winter. The heat is most easily measured in British thermal units (BTUs); or in comparison with other fuels. Different kinds of wood in identical volumes produce different amounts of heat. The denser the wood, the more material there is to burn. Therefore you'll end up paying more for a lighter-density wood than for a heavier-density one. The accompanying chart shows the densities of various kinds of wood, and also compares the amount of heat available in a cord of wood with the equivalent amount of other fuels (oil, coal, gas, and electricity) needed to produce the same amount of heat. On this chart, heat is measured in BTUs.

How Wood is Sold

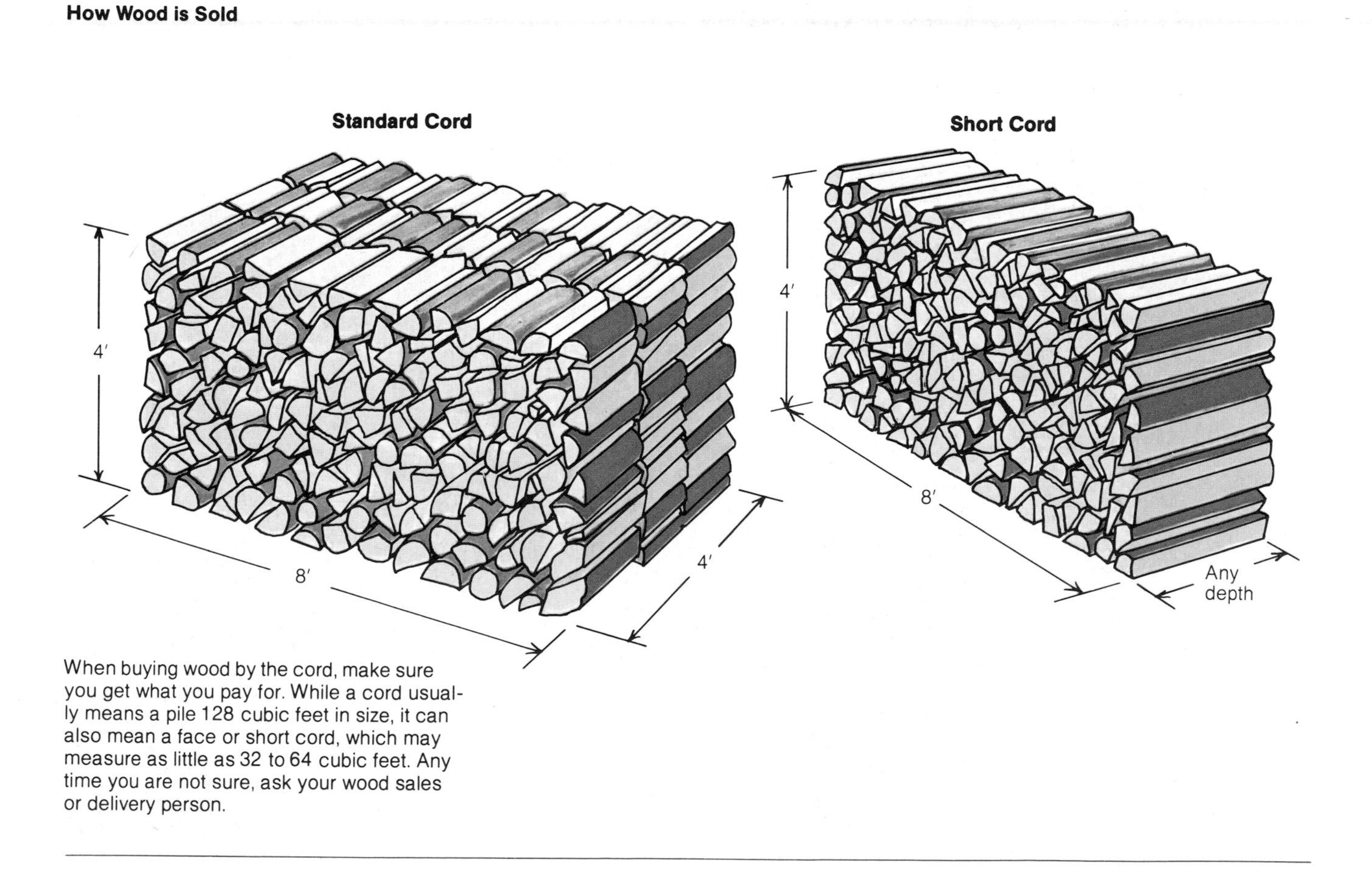

When buying wood by the cord, make sure you get what you pay for. While a cord usually means a pile 128 cubic feet in size, it can also mean a face or short cord, which may measure as little as 32 to 64 cubic feet. Any time you are not sure, ask your wood sales or delivery person.

Wood has always been used as a heating fuel, but with the advent of other inexpensive and accessible fuels, its popularity declined. However, when the prices of these other energy sources began to increase, wood received renewed attention. People have begun to seek out more information about cutting, buying and burning wood for use as a primary or secondary source of home heating. You will find the answers to your questions about wood as heating fuel on these and the following eleven pages.

Wood Densities and Fuel Equivalents

Species (Wood)	Weight (pounds per cord)		Available Heat/Cord (Millions of BTUs) at 7000 BTU/lb	Equivalent Heating				Heat	Remarks
	Green	Air-dry (20% moisture)		Gals Fuel Oil	Therms Natural Gas	Kwh Electricity	Gals LP Gas		
Ash	3840	3440	24.1	159	206	4234	226	High	Moderately easy to split. Will burn while still green.
Aspen	3440	2160	15.1	100	130	2659	142	Low	Very easy to split. Use for kindling only.
Beech	4320	3760	26.3	174	226	4628	247	High	Moderately difficult to split. Seasons best if split.
Paper Birch	3800	3040	21.3	140	182	3742	200	Med	Moderately easy to split. May rot if not split.
Yellow Birch	4560	3680	25.8	170	221	4530	242	High	Moderately easy to split. May rot if not split.
Elm	4320	2900	20.3	134	174	3570	190	Med	Very difficult to split.
Hickory	5040	4240	29.7	196	254	5219	278	High	Moderately difficult to split. Light sparks and smoke.
Soft Maple	4000	3200	22.4	148	192	3939	210	Med	Moderately easy to split. Light sparks and smoke.
Hard Maple	4480	3680	25.8	170	221	4530	242	High	Moderately difficult to split. Light sparks and smoke.
Red Oak	5120	3680	25.8	170	221	4530	242	High	Moderately easy to split.
White Oak	5040	3920	27.4	181	235	4825	257	High	Moderately difficult to split. Light sparks and smoke.
White Pine	2880	2080	14.6	96	125	2560	136	Low	Very easy to split. Kindling best use.

Cutting Your Own Wood

If you cut your own wood instead of buying it, do so six to nine months before you expect to burn it, to give it time to season. Cut the wood in spring and fall, when the weather is pleasant and insects are dormant.

The preferred tools for cutting wood are the double-bitted axe and the chainsaw.

■ **Double-bitted axe:** A middle-of-the-road double-bitted axe weighs about three pounds; a light one weighs as little as two pounds; and a heavy one weighs as much as four pounds. Axe handles run from about 24 to 30 inches long. In contrast to a single-bitted axe, a double-bitted axe gives you two tools in one. You can keep one of the two blades honed extremely sharp for felling and limbing trees, and the other blade honed slightly blunter for cutting roots, slashing small branches from a fallen tree, or splitting logs.

In general, choose an axe whose heft is slightly heavy, and whose handle reaches to the top of your thigh or a little higher, when the axe is standing on its head. But if you plan to do extensive chopping—felling trees or cutting through logs the size of a telephone pole for instance—buy the heaviest axe with the longest handle you can swing in a comfortable rhythm. The heaviness and length combined will let the axe do a large portion of the work, and allow you to swing it for a maximum amount of time with a minimum of effort and discomfort.

Technique is important when you're swinging an axe. Take a comfortable position in front of your wood and grasp the axe a couple of inches up from the end of the handle—*not* with one hand toward the heel and the other halfway toward the head. Look at the spot you want to strike; place the blade there; swing the axe back and up; and let it fall to its target. After a few cuts—when you've discovered that you don't have to *drive* the axe down, but that you only have to bring it up and guide it the rest of the way—you'll begin to find an easy, natural rhythm that will allow you to cut far more wood in any given period of time, and cut for far longer, than you can if you *work* at it.

All the axe lore in the world won't help you if your blade is dull, however. Not only will a sharp cutting edge make your job easier, it will also make your job as safe as possible, since it is the dull blade that glances off the log, and the sharp one that bites.

To sharpen an axe use a file or rasp to smooth the inevitable nicks, coat the edge of the blade with a drop or two of oil, and use a hand-held hone or sharpening stone that has at least two grades of grain for coarse and fine honing. Move the stone in a circular motion along the blade edge, being sure to maintain the original bevel. This is not a complicated job, but it takes time and more effort than the new wood chopper may expect. Take the time and make the effort; the energy you save will be your own.

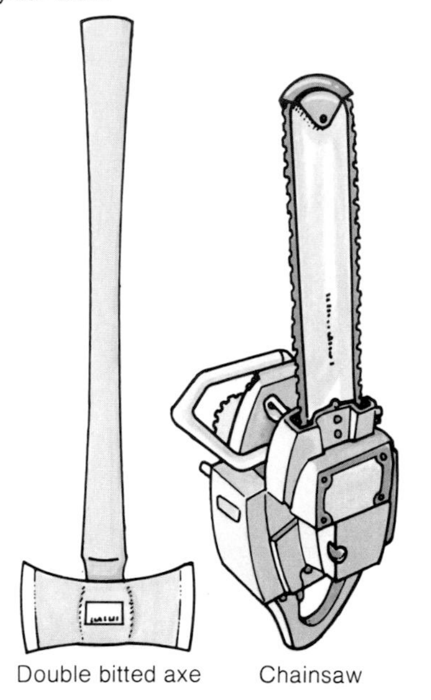

Double bitted axe   Chainsaw

■ **Chainsaw:** Chainsaws range in size from about 6 to 15 pounds. If you plan to do a lot of cutting, you can justify buying a heavier chainsaw, since it will make your work go much faster and easier. Of course, heavier saws are harder to hold and require more strength to use. If you expect to cut only one cord or less per year, a smaller, lighter saw will do just fine. The light ones are relatively easy to use, and they will take down small trees.

Chainsaws come in both electric- and gas-powered models. An electric saw is less expensive, but you may need to buy a portable generator to go with it, and the generator will make the cost considerably greater than that of a gasoline saw. Electric saws tend to be lighter, and are more useful for smaller jobs.

Chainsaws cost anywhere from $80 to $500, depending on the size, quality, manufacturer, and where you buy it. It's best to buy your saw new rather than used—that way you can be certain what kind of upkeep it receives from the beginning. You may prefer a particular brand of saw, especially if you've read reports on the various brands or have tried out a variety of saws. But if your winter heating depends to any degree on your ability to cut wood, buy a saw that can be serviced locally.

The chainsaw is a powerful tool, and it can be extremely dangerous if it is mishandled. So always work slowly, mindfully, and carefully, and always follow the manufacturer's instructions for use, care, and maintenance.

Felling Trees

There's a certain economic appeal to the idea of free firewood, and a romantic appeal to the idea of cutting your own. But don't let these notions blind you to the reality—this is not a sport for the faint-hearted. If you're going to attack anything larger than a 12- or 15-inch-diameter tree, or if you plan to hew a small grove in a day, you are going to *earn* your satisfaction, whether you use an axe or a chainsaw. You can, however, make your work easier and safer by following some simple precautions.

1. Wear **clothing** that fits just right—loose enough to keep your limbs free, yet tight enough not to get caught on branches or your own tools.
2. Wear **goggles** or an **eye shield**.
3. Wear a **hard hat**. Dead limbs that fall from trees while you are beneath them are called "widow makers."
4. Wear **ear plugs** if you're working with a chainsaw.
5. Wear heavy **boots** with steel-capped toes and non-skid soles, and try to keep them out of the way of rolling and falling logs.
6. Have **company**—always cut with another person present. If you've never felled trees before, choose a partner who has.

■ **How to fell a tree:** First make a wedge-shaped cut on the side of the trunk *toward* which you expect the tree to fall. Make a second cut on the other side of the tree slightly above the apex of the wedge. As the second cut nears the first, the tree will fall pivoting on the "hinge" of wood left between the cuts. When the tree starts to fall, get away from the direction of the fall as quickly as possible. Falling trees are not always predictable, but they tend to leap up and back when they hit the ground.

After the tree has fallen and you have trimmed off the crown and small branches, **limb** the tree by removing the larger branches. Then cut the boughs and trunk into firebox-size pieces. This process is known as **bucking**.

You should buck your tree as soon after felling it as possible, because if it lies around on the ground for very long it will take up moisture and begin to rot. For the same reason, you should split and stack your logs for seasoning as soon as possible after bucking.

■ **Splitting:** Splitting can be a lot of fun (unless you've felled an elm or some other split-resistant tree), and is really what most people have in mind when they envision cutting their own firewood.

In general, split between knots, rather than through them; along cracks, if there are any; green wood (it will split more easily than seasoned wood); light, straight-grained wood (it will split more easily than heavy wood with a wavering grain); and frozen wood (it will split more easily than unfrozen wood). If you've felled and bucked in late autumn, you might want to stack your unsplit wood, and simply split it as you need it after the weather turns freezing. Don't bother trying to split logs where large branches grew from the trunk—they'll almost never split, and in any case they will be too wide for use in most fireboxes.

Large, thick logs will be difficult to split by axe, and may require the use of a maul, which is like a blunt-tipped axe. Because of the shape and extra weight of the head, mauls force wood apart rather than cutting it.

Use some sort of chopping block when you split wood—the ground itself will dull your axe faster than you can chop wood. An old stump is excellent for this purpose if it is fairly low (about 15 to 20 inches high). Stand the log to be split on the block and examine it to be sure you won't be chopping through knots, which are cross-grained to your cutting angle, and very hard to split. Then split off the edges of the log into four or more pieces, leaving the barkless center. When this core is small enough, split it straight down the center. If your pieces get very small, just put them in the kindling bin.

■ **Stacking:** Once you've split your logs, stack them to dry out, or season. Obviously, you'll make it easy for yourself if you stack your wood near the door you'll be using to carry it to your firebox. Just make sure to keep it separate from the exterior walls of your house; insects are likely to be inhabiting the split logs. If you can find a small rise on which to locate your pile, do so—the ground underneath it will stay dryer than would the ground under a depression, which might collect water. Exposure to air and sun will speed the seasoning process.

Split wood will dry fastest in a **chimney** formation. Place two logs parallel to each other and less than one log's-width apart. Then put two more logs on top of the first two, but set them at right angles. Continue with the next two at right angles again (paralleling the formation of the first two), and so on up the stack. Ordinary parallel or cross-stacking occupies less space than chimney stacking, but it doesn't let air circulate as freely among the logs. Don't make a chimney stack too high; it might topple in a strong wind. Ideally, you will let your wood dry for a year, but you can burn it successfully before that.

Once your stacked wood is almost dry, cover it to protect it against snow, rain, and ice. A tarp or a sheet of heavy plastic works well. Don't make an airtight seal—air should be able to continue to circulate, even when the woodpile is dry.

How Wood Burns

It will take no more than five to ten cords of good firewood per year to completely heat the average American house. The exact figure will depend on such factors as climate; size, shape, and heat-retaining ability of the structure; and the efficiency of the wood-burning unit. In evaluating this last consideration, it helps to have some idea of the process by which wood becomes heat energy.

The efficiency of a wood-burning unit is a measure of its ability to produce heat from wood. Such a unit is rarely 100 percent efficient—not only because heat disappears up the chimney, but also because it is difficult to burn any wood completely.

Wood is a complex fuel whose chemical makeup changes as it burns. It is made up principally of carbon, hydrogen, oxygen, and a few trace elements, all of which participate in the burning process in different ways and at different times.

Wood undergoes three important transformations when it burns:

1. The water boils off;

2. The wood fiber breaks down, forming charcoal and emitting such volatile gases as hydrogen and carbon monoxide. The simpler wood fibers burn at about 750° F, but the gases require temperatures around 1100°F. These gases must be burned for wood to be an efficient fuel, since they provide 50 to 60 percent of the wood's heat; and

3. The charcoal burns.

Ordinarily, all three burning phases happen at once. Thus a single small fire may range in temperature from about 300°F up through 1100°F, or even hotter.

Not only does fire need oxygen in order to burn, but to heat efficiently it must have the right *amount* of oxygen, and at the right *times*. If it has too little oxygen, it will smolder and may die;' if it has too much, it will become a fast-burning bonfire, losing most of its heat up the chimney. The most useful wood-burning heating unit will let you control the rate at which air mixes into the fire. Then you can have a roaring bonfire when you enter a cold room, but you can also leave a grate of coals smoldering overnight so that you can start a new fire easily in the morning. And for those long winter evenings, you can maintain a slow and steady burn for maximum heat over the longest possible period of time.

What kind of wood-burning unit will give you maximum efficiency? Unfortunately, it isn't the standard, homey fireplace. Despite our romantic attachments to fires crackling in the fireplace, we have to deal with some unromantic facts. Let's take a look at the options.

Fireplaces

■ **Masonry fireplaces:** Houses built 50 to 75 years ago were designed to take advantage of a centrally located fireplace with its solid brick, heat-retentive chimney reaching up several stories through the center of the house. Today, however, fireplaces are largely decorative, are built into outside walls, and have a clay flue with 6 to 12 inches of space between the flue and the brick facing; therefore, the heat never really penetrates the brick, and what heat is retained is lost to the outside. Consequently, today's fireplace makes for a net heat loss in almost every case, despite the traditional picture of a family gathered happily in front of a blaze crackling in the hearth. The combustion efficiency is only about 10 percent, which means that 90 percent of the heat produced goes up the chimney. This unused heat is known as **stack loss**.

■ **Freestanding fireplaces:** A freestanding fireplace is basically a hood made of sheetmetal hung over some sort of grate, with a pipe or chimney to duct smoke away. The metal hood heats evenly and radiates that heat in all directions; compared with other fireplaces, these units have a relatively high combustion efficiency, particularly when set in the middle of the room.

■ **Metal fireplaces:** Another type of fireplace is the prefabricated metal variety that can be built into a wall or an exist-

The Chimney Effect

In a conventional masonry fireplace, most of the heat produced by burning wood is lost up the chimney or into the surrounding wall and floor. A small amount of heat circulates around the room, and another small amount radiates out to the persons or objects nearby. Cold air is drawn in from outside to replace the warmed inside air that escapes up the stack, making the room even cooler.

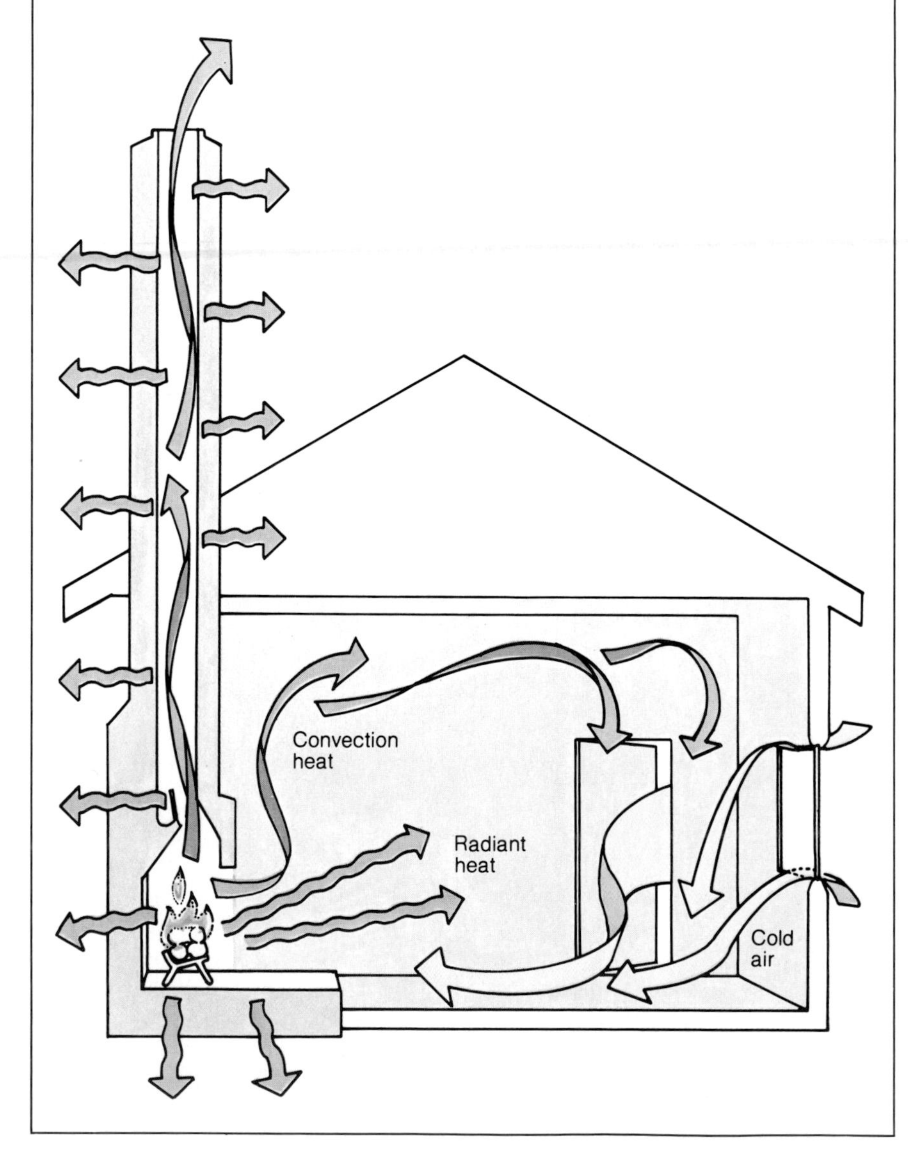

ing fireplace. While it is easy to install, not very expensive, and sometimes offers an optional forced-air, electric-powered fan that will improve its performance, the cost of such a unit is out of proportion to the heat it provides. Like the masonry fireplace, it is only about 10 percent efficient.

Even when there is no fire, a fireplace can cause a house to lose heat. If your fireplace has no damper, or if it does but you leave the damper open, your house has an air leak—one that may account for 20 percent of the air heated by an ordinary furnace every hour.

■ **Improving fireplace efficiency:** There are ways you can improve the efficiency of your fireplace. You can purchase and install a damper, if you don't already have one, which will minimize heat loss when no fire is burning.

You can also buy glass doors to surround the hearth. These will allow you to enjoy the fires while saving you the heat your furnace has worked so hard to provide (although they will reduce the usable *radiant* heat from the fire).

If you can duct outside air directly into your fireplace, you will significantly reduce the amount of infiltration it creates, and thereby reduce the potency of the chimney effect when the fire burns. If you have two fireplaces, one on an outside wall and one on an inside wall, consider blocking off the one on the outside wall.

Because a fire consumes oxygen, the more the open-hearth fire burns, the more air it must draw from someplace. If you haven't weatherized your house, a fire in your fireplace will draw enormous quantities of unheated air in from outside, through the walls, windows, and various other chinks and holes. You can easily make a bright, cheery fire in a traditional fireplace, but it will devour wood voraciously.

Even if you *have* weatherized your house it is still subject to a certain amount of infiltration; and an open fire draws cold air indoors faster than it would enter your house if the fireplace weren't in use. But whether your house is weatherized or not, as soon as the wood burns and creates heat, most of the heated air rises straight out your chimney. As it rises, it draws in *more* cold air—a cycle described as the **chimney effect**.

Dampers

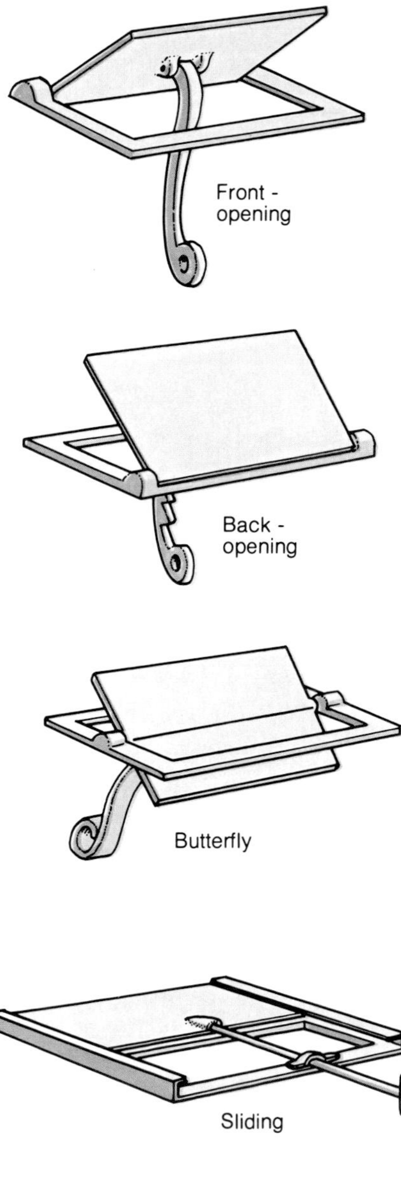

Ducted Circulating Fireplace

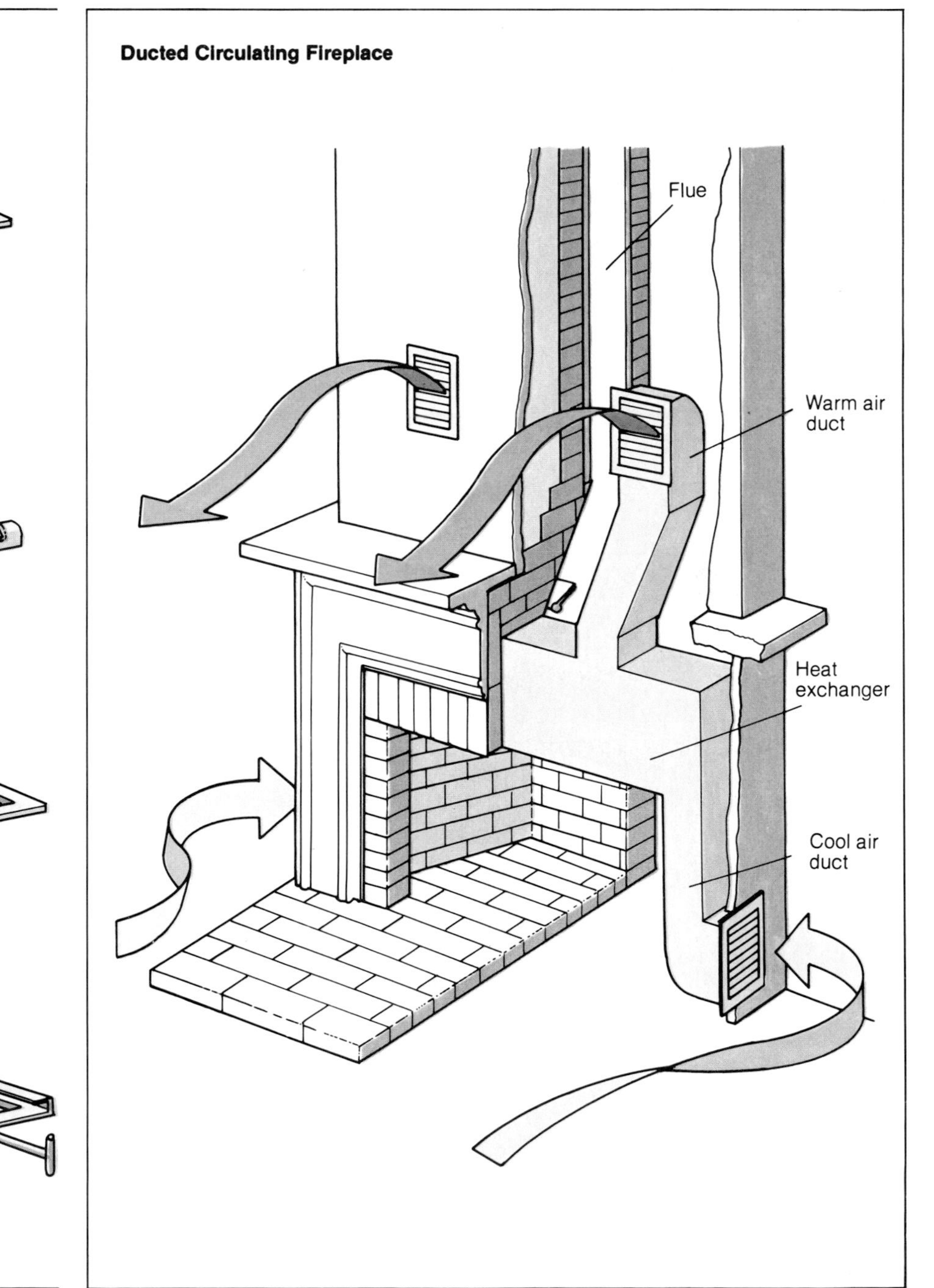

Heat Exchangers

Another way to gain heat from your fireplace is to buy or build a heat exchanger. Not to be confused with the heat exchanger inside a furnace or a solar water heater, this instrument is essentially a grate to hold your burning logs in a position that forces the heated air back out to you.

To build a heat exchanger, use lengths of 1-inch iron pipe, curved to rise from a position flush with the floor of your firebox until they point out into the room. The exact length of these pieces of pipe, as well as the number you will need, depends on the size of your fireplace and on the size of the logs you burn in it; however, they should be spaced at about 4-inch intervals. The top ends of the pipes should extend a bit beyond the fireplace.

Attach the curved pipes at the bottom to straight sections of the same kind of pipe, using ordinary pipe elbows. Weld the straight pieces of pipe to a plain iron fireplace grate, or to a few more pieces of pipe, bent to form legs. When a fire blazes in the heat-exchanger grate, cold air gets sucked into the open pipes at the bottom, where it's warmed by the fire and then forced out the top grate openings and into the room. In principle, this is a tiny chimney-effect. If you are venting cold air from the outside into your fireplace at the same time, this process will work even more effectively —the fire will get most of its oxygen from that air instead of from the air in your house that's already been heated. This device works best with fan-forced air. But with or without a fan, it will pump a certain amount of soot into your room, along with the heat.

Drawing on cool air from outside, or from the bottom of the room along the floor, a heat circulating fireplace returns warmed air out into the room through its ducts.

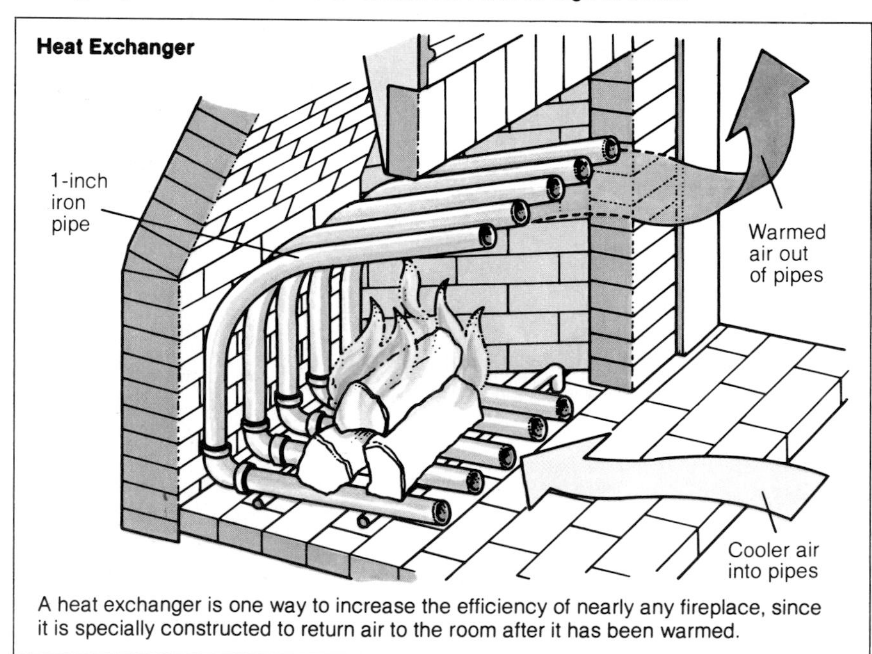

A heat exchanger is one way to increase the efficiency of nearly any fireplace, since it is specially constructed to return air to the room after it has been warmed.

Wood-Burning Stoves

■ **Pennsylvania fireplace:** In the 1740's, Ben Franklin, concerned about a local fuel shortage in Philadelphia and the inefficiency of existing fireplaces, invented the Pennsylvania fireplace. This device was essentially a metal box with a draft below that provided air for the fire to burn, and additional air that was circulated around the firebox before being forced back into the room by the construction of the instrument. When a metal top plate was added, Franklin's stove warmed the room both by convection and by radiation.

■ **The Franklin stove today:** Today's so-called Franklin stove is not so much a descendant of the original as it is a variation on the freestanding fireplace. Although it doesn't use the air-circulating system that made Franklin's fireplace remarkably efficient, it does use his folding or sliding doors, which allow the air flow into the firebox to be regulated easily. The efficiency of these stoves depends on tight seals all around the box, as Franklin himself noted. As a result, contemporary Franklin stoves cover a wide range of efficiency and inefficiency as heat-producing wood-burners.

Over the past few years, as Americans have demonstrated a renewed interest in burning wood for heat, the wood-burning stove has enjoyed a corresponding renaissance. Long-established manufacturers and dealers have prospered and grown, and dozens of new ones have emerged. Whatever your reasons for burning wood, a stove—and especially an airtight stove—is the best way to do it. Not only does it burn wood more efficiently and therefore heat more of your house longer for less money, a good stove actually costs less to install than a masonry fireplace, and is safer and more durable than a prefabricated metal one.

If you already have a fireplace, a wood-burning stove can be installed into this opening, even in an apartment. If you have no fireplace, and you are a renter, you can use a stove only if you can make the structural alterations necessary for proper venting. However, if you own your own home, you can do what you want, including making small holes in the roof.

Before you invest the $300 to $2000 that such a unit will cost (the price depends chiefly on size, weight, and ornamentation), you should know a little bit about how they work.

■ **How a wood stove works:** To some extent, all wood-burning stoves heat by *conduction* and *radiation.* But the most important mode of heat distribution for warming a space as large as a room or even a house is *convection.*

Conduction

Radiation

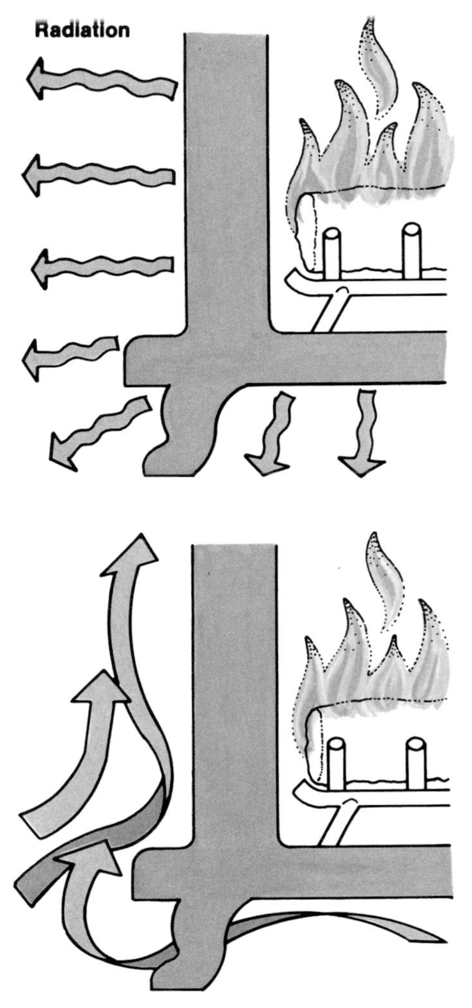

Convection

When wood burns in a stove, heat is conducted through the metal firebox, and radiates into the air around the stove. The warmed room air begins to rise up off the stove in the process of convection, heating further and further reaches of the surrounding area; as the warmed air rises, it is replaced by cooler air, which in turn is warmed. And so the cycle continues.

The airtight stove *(top)* allows you to control the influx of air to the firebox, and thereby to control the rate at which your fire burns. It allows for far greater efficiency in burning wood for heat than does the non-airtight stove *(below)*, which will burn with a merry blaze until it consumes its fuel.

Airtight Stoves

Essentially, all wood-burning stoves are either *airtight* or *non-airtight*. The airtight type is much more efficient because it lets the user control the air supply. As a result, you can restrict both the rate of combustion and the amount of heat lost in the burning process. Controlling these factors minimizes the expenditure of heat through the stack, which slows the chimney effect and lowers the amount of heat lost from the heated space. As a result, much less wood is consumed to produce a greater and more constant heat—the very definition of "efficiency" for a wood-burning unit.

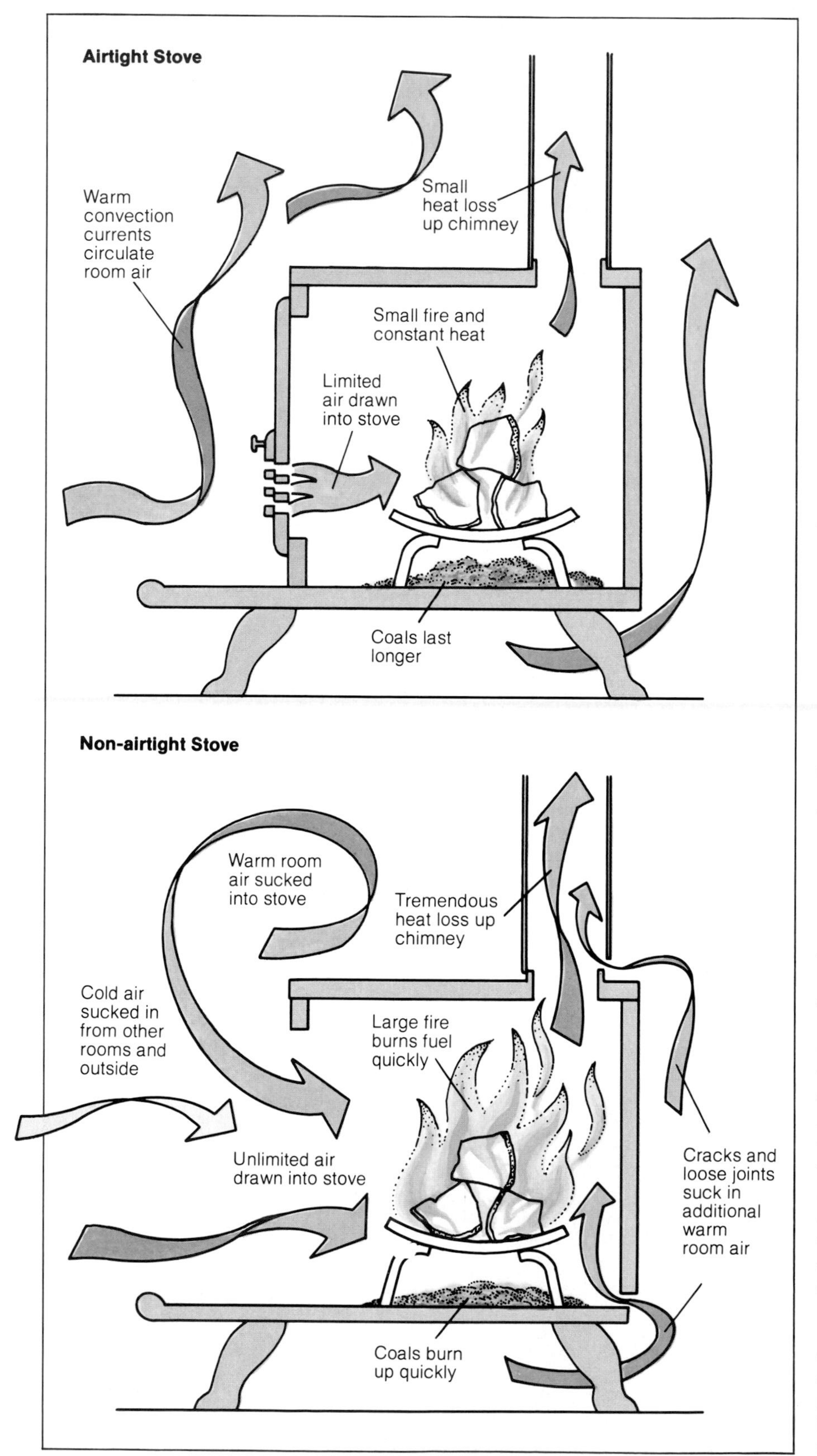

An airtight stove is also efficient in terms of how it distributes heat to the surrounding area. When a stove is not airtight (when it has cracks, loose joints, or other openings), heat is drawn back into the wood-burning unit and is lost up the stack. With an airtight unit, however, such drawback is minimal.

When all is working right—that is, when your airtight stove doesn't have to battle with infiltration from a poorly insulated house—it may take only minutes to begin heating your house in the morning. Then, once the air is hot, the convection cycle will keep it warm for as long as the stove is burning wood. And the convection cycle tends to create very even heat throughout your house.

Airtight stoves come in many forms, ranging from simple box-heaters and contemporary Franklin-type stoves to unique shapes that defy categorization. Most are made of steel. In stoves that are designed for maximum efficiency, gases from the burning wood are directed in specific patterns (draft configurations) to the hottest part of the stove where they are ignited, increasing the amount of heat.

■ **Combination stove:** One variation of an airtight stove is called a **combination** stove. With this type, the units can be either left open, providing a homey view of the fire that's burning inside, or closed for airtightness. This flexibility allows the rate at which wood is burned to be determined by air intake alone. As with ordinary airtight stoves, there is no need for a chimney damper.

Non-Airtight Stoves

The most popular non-airtight stoves are made of cast iron. The traditional parlor stove and potbelly stove have been manufactured for many years with small regard for airtightness.

Baffles, similar to those used in airtight stoves, have been introduced into some modern non-airtight designs. Although a few of the older non-airtight stoves were also built with baffles, many were not. If you want to use one of these older, nostalgia-producing artifacts today, try adding a damper to the duct carrying air to the fire—it will give you some measure of control over the wood's combustion (although you'll never have as much control as with an airtight stove). Fires in non-airtight stoves almost never keep their coals alive overnight, requiring a new fire and a new warm-up time each morning.

Further, the room housing the stove may be hot while the other rooms are only moderately warm. If this is your situation, you can remedy it to some extent by using a forced-air fan (with the intake near the heat source), or a series of fans between rooms.

Buying a Wood-Burning Stove

If you haven't yet bought a stove, an airtight is the obvious choice. It can heat your entire house to a toasty temperature, cost-effectively (unless your house is poorly insulated, in which case your heat will go up in smoke); a non-airtight stove, although it may heat your whole house or just a room or two, will burn more wood and consume it less efficiently than an airtight one.

But whatever your choice, you can get most of the information you need at any store that sells wood-burning stoves. Here are a few things to keep in mind. A good wood-burning stove may last 20, 30, or 40 years, or even more, but few are guaranteed for longer than 5 years. If you are seriously considering buying a stove, choose your manufacturer with care.

Although some unapproved stoves are quite good, most of the best ones will bear a seal, sign or tag indicating that they have been tested and approved by a reputable independent testing laboratory or other organization. Particularly with an airtight stove, but also to some degree with a non-airtight one, the door should fit in a way that encourages the unit's maximum efficiency. *Look* at the unit you are thinking of buying. *Touch* it. If you are looking at a cast-iron stove, make sure that the seams are all well-sealed, and that the castings are relatively smooth, with few uneven places or rough bumps and gouges. If you are examining a steel stove, determine that there are no gaps on the welds that will turn into air leaks.

Whether made of steel or cast iron, the walls of a wood-burning stove should be $\frac{3}{16}$ inch or slightly thicker. If the walls are much thinner, steel stoves may warp and bend under prolonged intense heat, and cast iron will burn out more quickly. However, if the walls are much thicker, they may absorb more heat than necessary and thus retard the unit's ability to heat a room.

■ Where to place your stove: Some rooms make more appropriate wood-stove sites than others. The ideal site is a room that is central, close to the other rooms, and low-ceilinged. In general, the more open and accessible the stove-heat is to the other rooms, the better the convection throughout the entire house will be.

If you have a sprawling ranch-style house, you may want to consider putting a stove at either end of the house. If you have a two-story house, you might want to put it near the stairwell to help heat the upstairs rooms. You can even install floor vents if you want to. You should not put your stove in the basement, (unless that is what you want to heat) even though heat does rise. The radiant heat from the stove will stay in the basement and be of little use to your upstairs living spaces. Within a room, it is better to keep it away from windows, so that you do not lose the heat through the glass. If you put it by your existing fireplace, you will want to block the fireplace opening. (See next page).

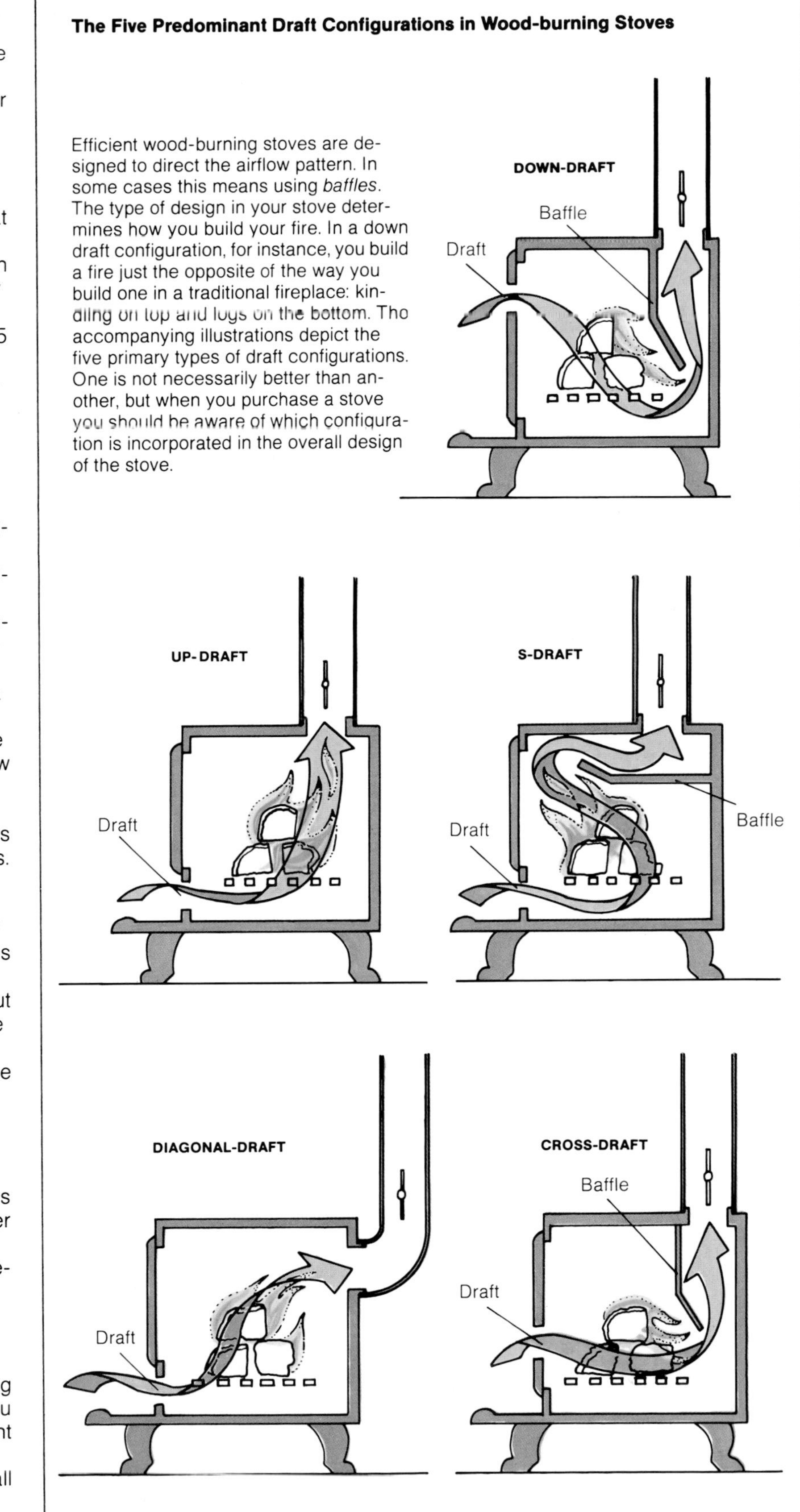

The Five Predominant Draft Configurations in Wood-burning Stoves

Efficient wood-burning stoves are designed to direct the airflow pattern. In some cases this means using *baffles*. The type of design in your stove determines how you build your fire. In a down draft configuration, for instance, you build a fire just the opposite of the way you build one in a traditional fireplace: kindling on top and logs on the bottom. The accompanying illustrations depict the five primary types of draft configurations. One is not necessarily better than another, but when you purchase a stove you should be aware of which configuration is incorporated in the overall design of the stove.

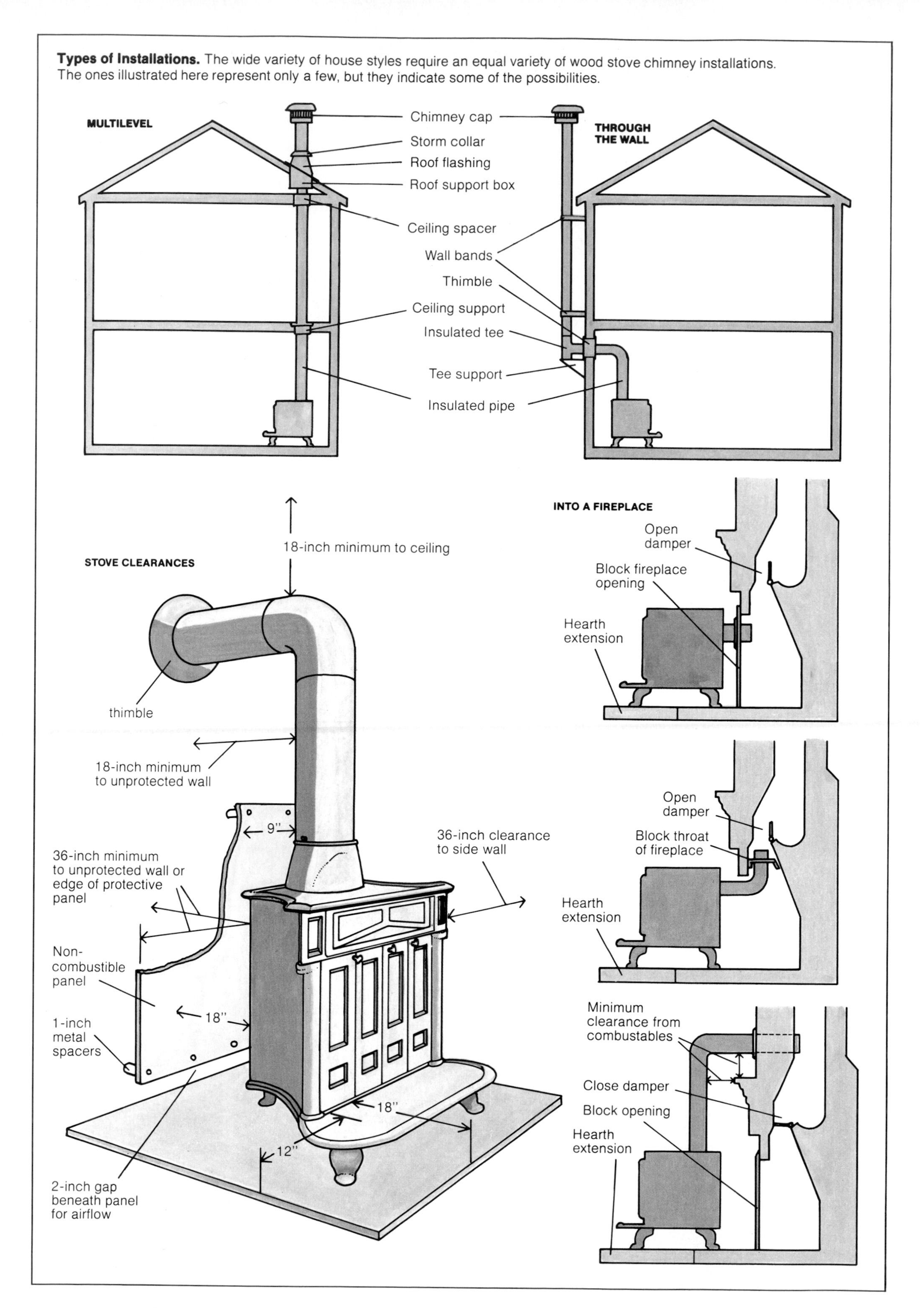
Types of Installations. The wide variety of house styles require an equal variety of wood stove chimney installations. The ones illustrated here represent only a few, but they indicate some of the possibilities.
MULTILEVEL
THROUGH THE WALL
Chimney cap
Storm collar
Roof flashing
Roof support box
Ceiling spacer
Wall bands
Thimble
Ceiling support
Insulated tee
Tee support
Insulated pipe
STOVE CLEARANCES
18-inch minimum to ceiling
thimble
18-inch minimum to unprotected wall
9"
36-inch minimum to unprotected wall or edge of protective panel
36-inch clearance to side wall
Non-combustible panel
18"
1-inch metal spacers
18"
12"
2-inch gap beneath panel for airflow
INTO A FIREPLACE
Open damper
Block fireplace opening
Hearth extension
Open damper
Block throat of fireplace
Hearth extension
Minimum clearance from combustables
Close damper
Block opening
Hearth extension

Installation

Some stove stores will install the unit for you; others won't. If not, the salesperson can usually refer you to a local independent contractor. The cost of such an installation, whether done through the store or not, will range between about $150 and $250 (depending on the difficulty of the installation), and will take anywhere from a few hours to a couple of days to complete. However, if you are reasonably handy and *very* careful—most stove-related hazards come from faulty installation, not the stoves themselves—you can install the stove yourself, venting it either through an existing fireplace or through the roof or ceiling. To some degree, installation procedures vary according to what type and brand of stove you buy, and what kind of ceiling and roof you have. You should discuss all such details with the salesperson.

In installing your unit, make sure to follow both the manufacturer's instructions and the building codes and ordinances that apply in your area. In general, if you keep the stove 36 inches or more away from any combustible wall, you shouldn't need additional fire protection on the stove's sides. Underneath the unit you should have *at least* 3/8 inch of noncombustible material, extending a foot or more on both sides and in back of the stove, and 1 1/2 feet in front. *These figures are minimal guidelines only*. If the manufacturer of your stove, or your local laws, require a greater leeway, follow those instructions. If you are in doubt about necessary space allowances, call your local city or county building inspector.

Creosote Buildup Can Cause a Fire in Your Stack

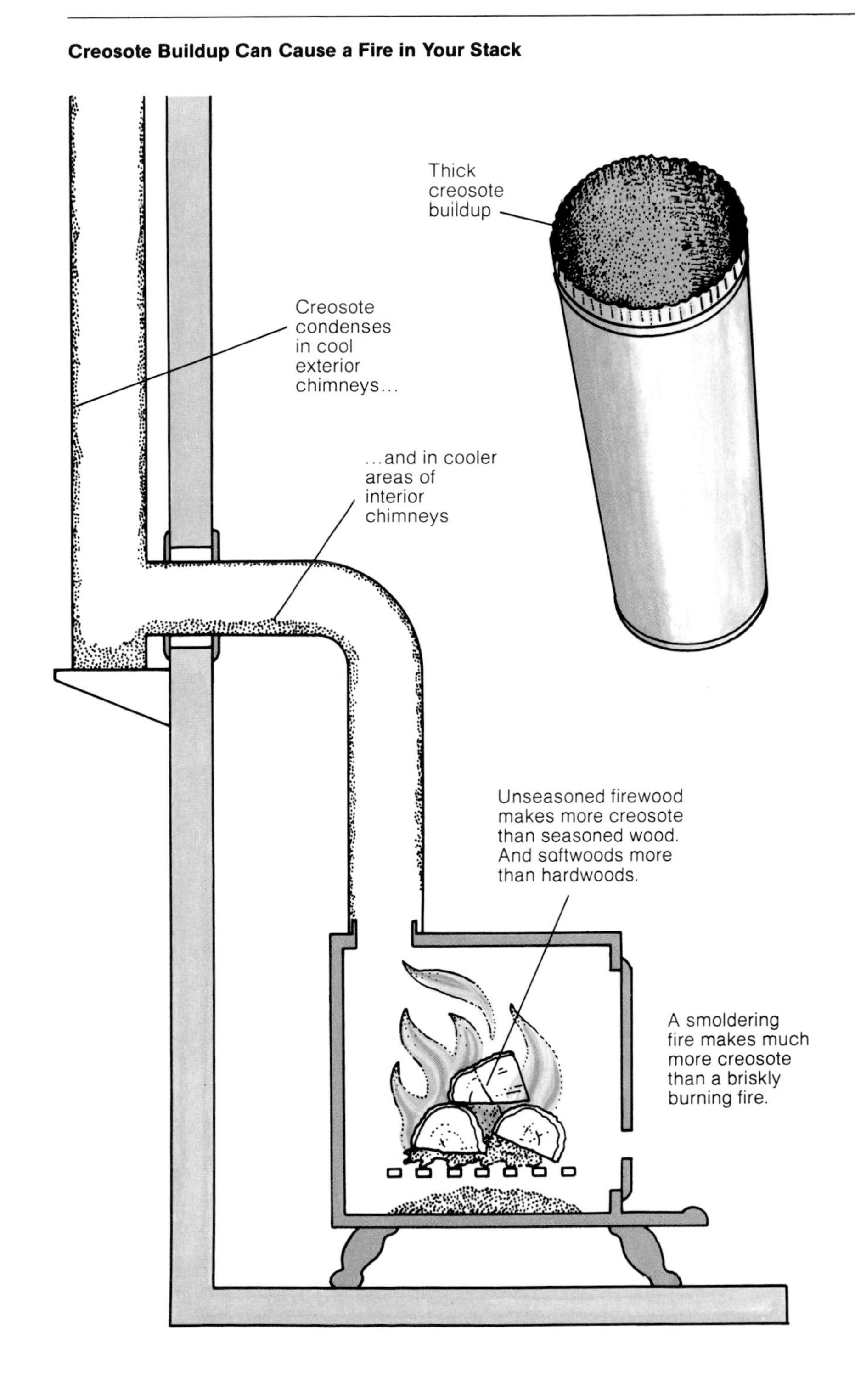

■ The importance of cleaning chimneys: With either a fireplace or woodstove, keeping your chimney clean is an important task when you burn wood, because creosote can build up in your stack and cause a fire in the chimney. This creosote is actually misnamed: it isn't the stuff used to preserve railroad ties, fence posts, and other outdoor woods—that's a coal derivative. The creosote we're concerned with here results from the process of burning wood.

Even the driest wood contains some moisture, and some woods contain a great deal. When wood is burned, and especially when it is burned with a low draft (as in an airtight stove left to smolder overnight), that moisture condenses against the insides of the relatively cool chimney. The steam that condenses there is not pure water vapor. It is pyroligneous acid, which contains small quantities of the volatile gases and particles of wood that have not burned completely. The lower the draft, the lower the temperature of the fire; therefore, the higher the proportion of such volatile matter in the steamy smoke rising in the flue.

As the acid condenses inside the chimney, it dribbles back toward the firebox. As it dribbles back, the remaining water is driven off by the heat, leaving a thick, brownish paste inside the stack. This paste is the creosote we're concerned with. This concern stems from the fact that not only will creosote buildup restrict the available draft, but—far worse—it is very flammable. When it builds up in large enough quantities it can cause a dangerous chimney fire.

Burning dry wood or keeping your draft high will minimize creosote buildup, but it won't eliminate the problem. The only thing to do is to clean out your chimney periodically. And whenever you clean your chimney, you might as well clean your stove as well. How often you'll need to clean will depend on the quality of the wood you burn and how often you burn it. Climate also is a factor—in San Francisco, you'd probably clean your chimney once a year, but in New England you'd most likely do it as often as three times a year. But these figures are merely guidelines—always follow your manufacturer's recommendations and your local codes; and in the absence of either of those, ask advice from your dealer about the particular stove you own.

One way to clean your chimney is so easy that you don't even have to get your fingers dirty (most of the time). All you do is toss chemical salts onto your hot fire—salts such as sodium chloride, calcium chloride, or sodium chloride mixed with copper sulfate. The resulting smoke flakes the creosote from the inside walls of the chimney; then the creosote falls into the fire and burns. You don't have to apply a chimney brush more than once every half-dozen cleanings or so. These chemicals are available under various brand names in hardware stores, fireplace supply outlets, and wood-stove stores. However, because these salts can produce some chlorine compounds, some air pollution agencies may prohibit their use.

Of course, if you want to harken back to Dickensian days, you can be your own chimney sweep. It's a messy job, to be sure, but fairly straightforward.

First, close off the mouth of your fireplace with thick plastic. Tape the plastic around the opening securely. Cover your rugs, furniture, the cat, and anyone else who hangs out near the fireplace. Some soot and dust are going to filter in under almost any circumstances, no matter what you do.

Now cover yourself as well as possible with clothes you don't mind getting grimy. Wear a dustmask over your mouth and nose, and goggles over your eyes.

Tie a stiff steel brush—one that's either finned or made up of several wheels—to the end of a rope that's as long as your chimney is high, and weight down the end of the brush. Chimney brushes are generally available at wood-stove outlets, where you will also be able to get advice about this task from service-oriented professionals.

Climb onto your roof *carefully*, carrying the tool with you. Drop the weighted brush down the chimney all the way to the smoke shelf, and pull it back up, scraping the steel brush against the insides of the flue. Repeat this operation until there is very little new creosote coming up on the tines of your brush. If your chimney is already reasonably clean, you may have to dip the brush as few as five or six times; if the flue is really dirty, you'll be at the job considerably longer. Bring a large can or other container along with you to put the creosote in. You can use the same brush to clean the stovepipe sections; just take the sections down and take them outside first.

Of course, you don't *have* to sweep out your chimney yourself. As wood heating has regained some of its former popularity, the profession of chimney sweep has come back into its own. There are chimney sweeps in most urban areas of the United States. Ask your stove dealer for recommendations.

You Can Remove Creosote from Your Own Chimney

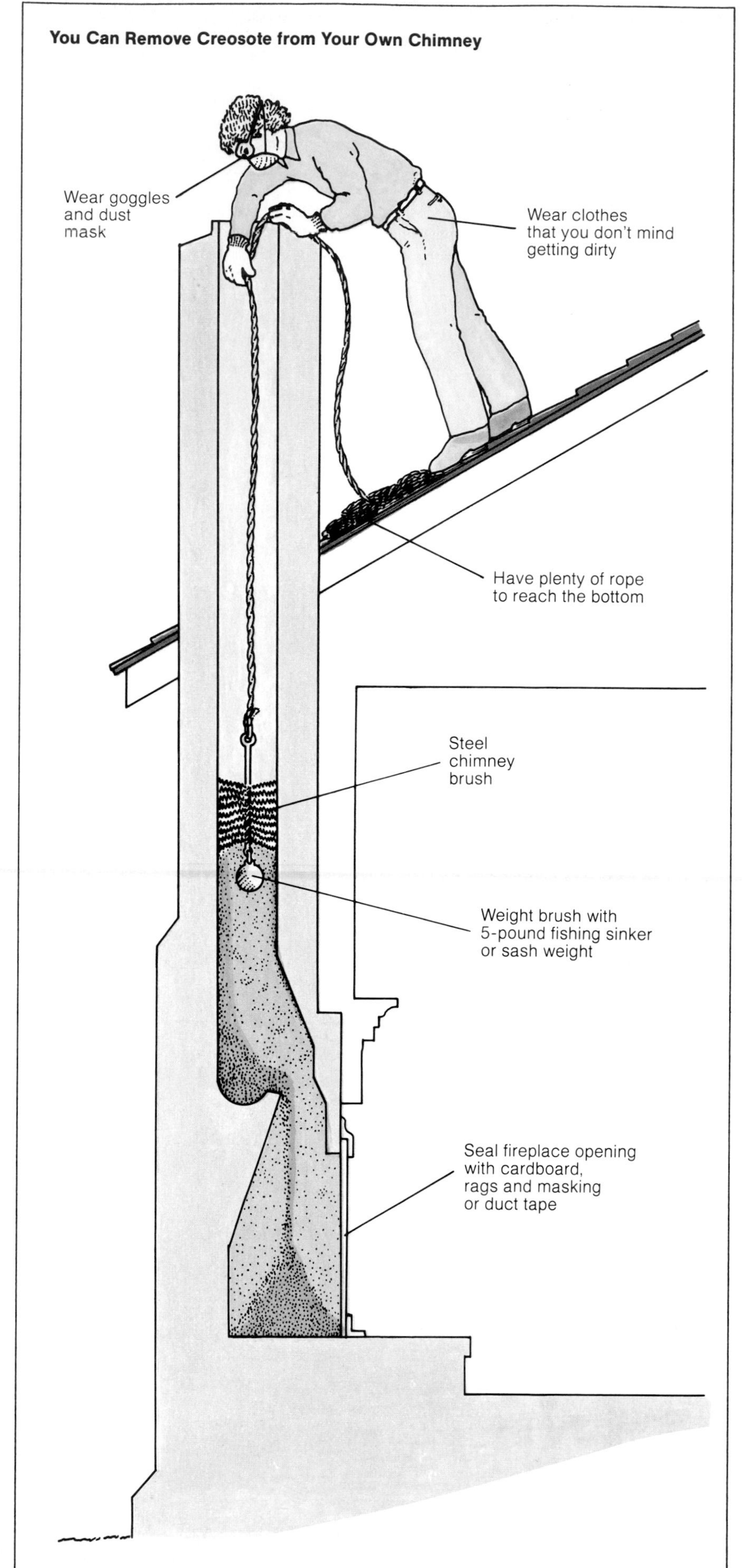

Water and Wind

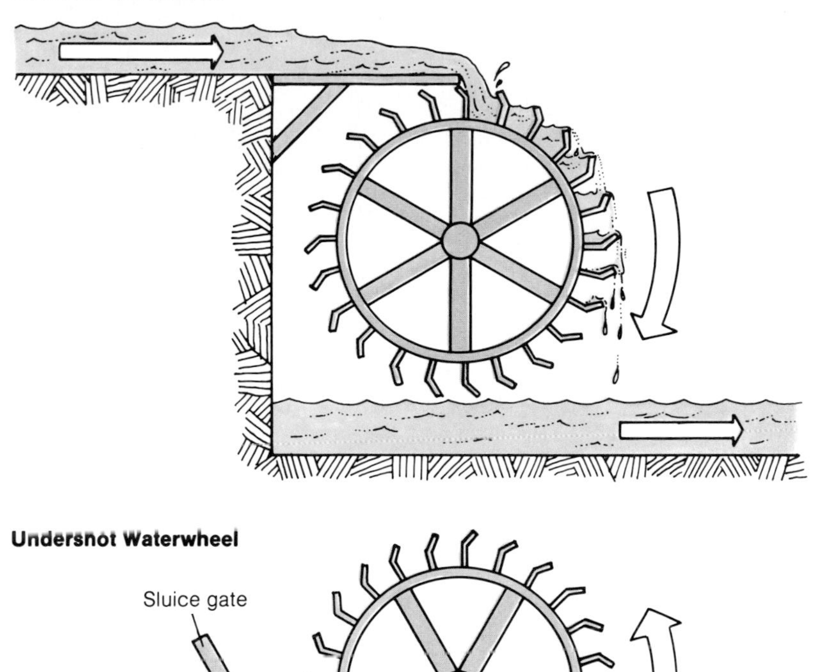

Overshot Waterwheel

If you had just landed on these shores a few hundred years ago, water and wind would have been the energy sources you could most readily see and use (apart from wood and—in a passive fashion—the sun). You'd have needed only a few tools and a modest technology, and you would already have been familiar with their ancient, worldwide applications.

Of all the known energy alternatives we have not yet discussed, water and wind are the two that individuals, as well as large concerns, can use to generate electricity. However, although such a feat is possible, it is generally unrewarding, and it requires special precautions.

■ **Water:** Water is clean, renewable, and an extremely efficient source of power. Properly harnessed, 80 to 90 percent of the available energy in a moving stream can be used, compared with the 25 to 50 percent energy efficiencies of other major power sources.

Simple water wheels have been used to grind grain for at least three thousand years. More recently, massive multipurpose installations have provided irrigation and flood control; lakes and waterways in the form of dammed-up reservoirs; and about 12 percent of America's electric power. Encouraged to seek out other options by the rising costs of fossil fuels, a growing number of individuals, on their own and with the support of government and industry, now consider water to be a component of their long-term energy programs.

If you live in the heart of a major metropolis, you will have to wait for water power to come to you. But if you live on or near even a small vigorous stream; if you like the idea of energy-independence; if you enjoy working with your hands; and if you have a few thousand dollars to invest in your dreams, you can provide yourself with an independent hydroelectric plant that will run anything from a few lightbulbs and your kitchen clock to your most complex electric appliances.

Unless you're an experienced handyperson, building and installing your own waterwheel is going to be very hard and possibly dangerous work; and when you've finished, the chances are excellent that you will find it would have been cheaper to tie into the existing energy grid than to make your own. Nevertheless, some people have constructed wheels that generate enough electricity to provide an excess, which the proud wheel-owners then sell to the local utility, thereby amortizing their expenses on the project more quickly than they could have otherwise.

All hydroelectric plants need moving water. Whether you are already on a piece of land with a stream, or a river, or whether you are looking for watercourse acreage to buy, consult with a lawyer or your local Department of Agriculture office early in the planning stages to familiarize yourself with all laws and regulations governing water power.

Assuming that you have found the right watercourse, and that you can build a wheel where it flows, make sure that the water moves 12 months a year —if it dries up in summer or freezes solid in winter, you are up the creek without any power. However, if the water does run all year, even if the stream is small, you will probably have enough of what you need to achieve minimal electrical generation.

Water power is measured by its **flow** (the volume and speed of the stream) and its **head** (the distance the water falls from above to the exit of your power plant). **Potential energy**, stored in the water supply, becomes **kinetic energy** as the water runs down the sluice that channels it to your wheel; and it becomes **work** when the water strikes and drives that wheel.

Your wheel's efficiency is determined by how much of the water's movement is consumed in turning your wheel. The ideal power plant directs a fast mass of water against the blades of the wheel with exactly enough impact to transfer all of the water's kinetic energy to the wheel and turn it. Efficiency of 100 percent is achieved when the water then simply stops.

In order to provide enough water with enough head that moves often enough to make a waterwheel workable, you need some form of water reservoir and some form of precipice (frequently these are combined by use of a dam). The best situation is for the water supply to be backed up into a natural holding-tank such as a swamp. Ideally, the water drains from the swamp at its narrowest point, raising the velocity of flow just before the water drops over a waterfall, to provide you with maximum head at minimum expense. Otherwise, you may be faced with some serious work-dredging your reservoir and building your dam—before you are ready to go to work on your waterwheel.

The simplest type of waterwheel is the **overshot** wheel, which requires neither fancy equipment nor delicate balancing acts to build. Water is carried through a sluice above the wheel; a gate regulates the amount of water entering each bucket, or striking each blade, on the wheel just as it reaches its peak; and the buckets hold their water as far down the wheel as possible before turning and spilling it, in order to produce the largest possible energy transfer and power gains.

A slightly more complicated version is the **undershot** wheel, in which only the lowest buckets or blades are struck by the moving water. The impact of the water, combined with the weight of the water itself, moves the wheel.

In our sophisticated times, these simple waterwheels have proven inefficient for the large-scale generation of hydroelectric power. **Turbines**, originally developed during the Industrial Revolution, can be highly efficient, but they require precise engineering. In a turbine, the sluice directs the flow of water to strike the turning wheel's blades at a specific, very exact angle to produce high turning-speeds and, consequently, large quantities of electric power.

Obviously, if you plan to dip into water power at all, you will need either some training or a good deal of trained assistance. For information ranging from legal issues to technology to the names of contractors, call your local US Department of Agriculture Resources Conservation District. But stay on top of your project—otherwise you may find yourself with a wheel that doesn't turn, suspended beneath a dam that doesn't hold, over a stream that no longer flows, going nowhere and producing nothing on a project that will cost you several thousand dollars at the very least.

■ **Wind:** Like water, wind is clean, renewable, and potentially efficient. Like waterwheels, windmills have been used to grind grain and to pump water for thousands of years. And like water power, wind power has received a burst of technological attention in recent years.

Just as you can harness water power, so can you harness wind power. But again as with water power, the expense of constructing the apparatus for a small-scale operation is unlikely to be recouped in any great hurry. In fact, unless your present electric bill exceeds $100 per month, the windspeeds around your house average more than 10 miles per hour, and you have a bare minimum of $7,500 to blow away, your personal pursuit of wind power should be accompanied by a deep religious commitment.

Windmills have a prodigious number of application problems. First, wind energy is highly site-specific. Just because the wind is blowing *here* and *now* does not mean that it will blow *there* and *then*, even though "now," "here," "there," and "then" may be only a few feet or minutes apart.

Second, any wind-catcher must be located well above its surroundings, such as buildings, hills, and trees.

Third, wind is a very intermittent energy source. The wind—especially a strong one—does not blow regularly, in most places.

Wind power is most useful if the work it performs can be stored. Typically, wind power is collected for electrical use with a DC generator, and transformed to AC for storage in batteries. But it can also be tied into a public utility grid to be fed into the larger system when an excess is available.

Wind Electric-power-generating System

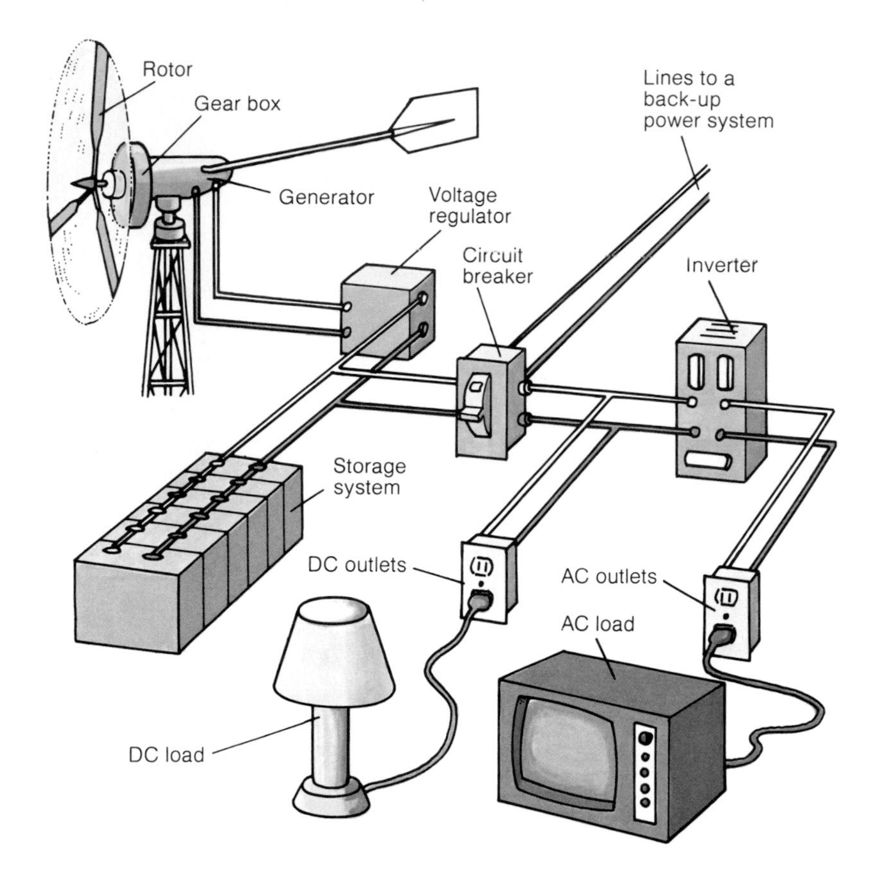

Most electric utilities in this country charge about 4¢ to 6¢ per kilowatt hour. Since your generating cost is likely to be as much as 20 times this amount, you will not come out the winner in this exchange. However, you may be able to cut your losses.

Any electricity-generating windmill is an integrated system, composed of propellers, a generator, a voltage regulator, inverter and some sort of storage. If you are keen on becoming intimate with your system and plan to do only a nominal amount of work with your self-generated electricity, you may be able to put together a system from components on your own. But if you plan to do a great deal of work with your electricity —if, for instance, you plan to operate not just your light bulbs and TV but also your washer, dryer, and refrigerator—then there's no point in buying anything less than a complete wind-generator system.

In a wind generator, the propellers are turned directly by the wind. Because they ordinarily turn too slowly to ensure the high rotational speeds necessary to run the generator, they usually are connected to it through step-up gears. The generator turns the wind energy into electricity, which is monitored by the voltage regulator to provide constant electricity to the batteries, which store the electricity for use on windless days.

If you've decided to convert to wind power, you must be knowledgeable about a variety of factors about your own energy consumption as well as about the hardware available.

1. Know your annual and seasonal consumption, and know which uses of electricity you are willing to do without, if necessary.

2. Know the annual and seasonal windspeed averages where you plan to erect your windmill.

3. Know that your wind access cannot be blocked by a new highrise or a growing stand of trees.

4. Go out to see existing windmills; talk to their owners, the people who installed them, the people who sold them, and the people who manufactured them, if possible. Find out what, among the available hardware, truly suits your needs.

5. Assume that you will need more electricity than you think you will, and that your windmill will produce less than you think it will. These pessimistic forecasts will make you happier in the long run.

6. Provide yourself with a backup power source for emergency situations.

7. Remember that you are dealing with far more power than is necessary to kill a person, and be careful. (This is true whether you are building the windmill yourself or having it built and/or installed for you.)

INDEX